MATHEMATICAL APPLICATIONS AND MODELLING

Yearbook 2010
Association of Mathematics Educators

MATHEMATICAL APPLICATIONS AND MODELLING

Yearbook 2010
Association of Mathematics Educators

editors

Berinderjeet Kaur

Jaguthsing Dindyal

National Institute of Education, Singapore

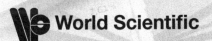

Published by

World Scientific Publishing Co. Pte. Ltd.
5 Toh Tuck Link, Singapore 596224
USA office: 27 Warren Street, Suite 401-402, Hackensack, NJ 07601
UK office: 57 Shelton Street, Covent Garden, London WC2H 9HE

British Library Cataloguing-in-Publication Data
A catalogue record for this book is available from the British Library.

MATHEMATICAL APPLICATIONS AND MODELLING
Yearbook 2010, Association of Mathematics Educators

Copyright © 2010 by World Scientific Publishing Co. Pte. Ltd.

All rights reserved. This book, or parts thereof, may not be reproduced in any form or by any means, electronic or mechanical, including photocopying, recording or any information storage and retrieval system now known or to be invented, without written permission from the Publisher.

For photocopying of material in this volume, please pay a copying fee through the Copyright Clearance Center, Inc., 222 Rosewood Drive, Danvers, MA 01923, USA. In this case permission to photocopy is not required from the publisher.

ISBN-13 978-981-4313-33-9
ISBN-13 978-981-4313-34-6 (pbk)

Printed in Singapore.

Contents

Part I	**Introduction**	1
Chapter 1	A Prelude to Mathematical Applications and Modelling in Singapore Schools Berinderjeet KAUR Jaguthsing DINDYAL	3
Part II	**Applications and Modelling in Primary School**	19
Chapter 2	Communities of Mathematical Inquiry to Support Engagement in Rich Tasks Glenda ANTHONY Roberta HUNTER	21
Chapter 3	Using ICT in Applications of Primary School Mathematics Barry KISSANE	40
Chapter 4	Application Problems in Primary School Mathematics YEO Kai Kow Joseph	63
Chapter 5	Collaborative Problem Solving as Modelling in the Primary Years of Schooling Judy ANDERSON	78
Chapter 6	Word Problems and Modelling in Primary School Mathematics Jaguthsing DINDYAL	94

Chapter 7	Mathematical Modelling in a PBL Setting for Pupils: Features and Task Design CHAN Chun Ming Eric	112
Chapter 8	Initial Experiences of Primary School Teachers with Mathematical Modelling NG Kit Ee Dawn	129
Part III	**Applications and Modelling in Secondary School**	**149**
Chapter 9	Why Study Mathematics? Applications of Mathematics in Our Daily Life Joseph Boon Wooi YEO	151
Chapter 10	Using ICT in Applications of Secondary School Mathematics Barry KISSANE	178
Chapter 11	Developing Pupils' Analysis and Interpretation of Graphs and Tables Using a *Five Step Framework* Marian KEMP	199
Chapter 12	Theoretical Approaches and Examples for Modelling in Mathematics Education Gabriele KAISER Christoph LEDERICH Verena RAU	219
Chapter 13	Mathematical Modelling in the Singapore Secondary School Mathematics Curriculum Gayatri BALAKRISHNAN YEN Yeen Peng Esther GOH Lung Eng	247

Chapter 14	Making Decisions with Mathematics TOH Tin Lam	258
Chapter 15	The Cable Drum – Description of a Challenging Mathematical Modelling Example and a Few Experiences Frank FÖRSTER Gabriele KAISER	276
Chapter 16	Implementing Applications and Modelling in Secondary School: Issues for Teaching and Learning Gloria STILLMAN	300

| **Part IV** | **Conclusion** | **323** |
| Chapter 17 | Mathematical Applications and Modelling: Concluding Comments
Jaguthsing DINDYAL
Berinderjeet KAUR | 325 |

Contributing Authors · 336

Part I

Introduction

Part 1

Introduction

Chapter 1

A Prelude to Mathematical Applications and Modelling in Singapore Schools

Berinderjeet KAUR   Jaguthsing DINDYAL

This chapter introduces the reader to the aims of mathematics education in Singapore and the framework of the school mathematics curriculum. It also sheds light on the intended curriculum, specifically the emphasis placed on applications and modelling in the framework since the last revision of the curriculum. For the past two decades, the framework has emphasised heuristics and thinking skills as problem-solving processes and teachers are comfortable teaching for problem solving and teaching about problem solving. Since 2007, teachers have been faced with the challenge to embrace the process: applications and modelling and expand their repertoire of means to nurture mathematical problem solvers. Finally, it introduces the reader to the chapters in the book, as it unfolds how each contributes to the theme of the book: Mathematical applications and modelling.

1 Introduction

In Singapore, mathematics is a compulsory subject for students from primary one till end of secondary schooling. Some students have to do it for 10 years while others for 11 years depending on their course of study at the secondary school. The mathematics curriculum adopts a spiral approach and has coherent aims. The aims stated specifically in the syllabus documents of the Ministry of Education (2006a, p. 5; 2006b, p. 1) are as follows:

- Acquire the necessary mathematical concepts and skills for everyday life and for continuous learning in mathematics and related disciplines.
- Develop the necessary process skills for the acquisition and application of mathematical concepts and skills.
- Develop the mathematical thinking and problem-solving skills and apply these skills to formulate and solve problems.
- Recognise and use connections among mathematical ideas, and between mathematics and other disciplines.
- Develop positive attitudes towards mathematics.
- Make effective use of a variety of mathematical tools (including information and communication technology tools) in their learning and application of mathematics.
- Produce imaginative and creative work arising from mathematical ideas.
- Develop the abilities to reason logically, communicate mathematically, and learn cooperatively and independently.

It is apparent that the aims provide a focus for the intended curriculum which is succinctly summarized in a framework, called the framework of the school mathematics curriculum. This framework summarises the essence of mathematics teaching and learning and serves as a guide for the implementation of the curriculum. The intended curriculum is presented as syllabus documents that outline the philosophy of the syllabus and objectives of the curriculum along with syllabus content for each level and course of study.

2 Framework of the School Mathematics Curriculum

The framework, shown in Figure 1, has mathematical problem solving as its primary goal. Five inter-related components, namely, *Concepts*, *Skills*, *Processes*, *Attitudes*, and *Metacognition*, contribute to the development of mathematical problem-solving ability. The framework developed in 1990, underwent revisions in 2000 and 2006. Table 1 shows the evolution of the component *Processes* in the framework from 1990 till the present. The 1990, 2000 and 2006 syllabuses were implemented in 1991, 2001 and 2007 respectively.

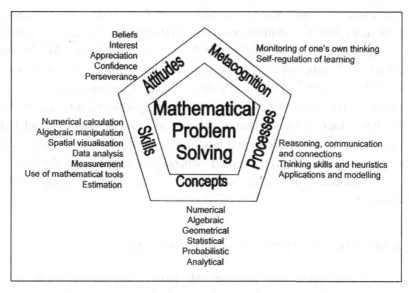

Figure 1. Framework of the school mathematics curriculum

Table 1
Processes: 1991 – present

Processes		
1991 - 2000	2001 - 2006	2007 - present
Heuristics	Heuristics	Reasoning, communication and connections
Deductive reasoning	Thinking skills	Applications and modelling
Inductive reasoning		Heuristics
		Thinking skills

From Table 1, it is apparent that the revised syllabuses of 2006 have placed emphasis on reasoning, communications and connections;

applications and modelling in addition to heuristics and thinking skills as processes that should pervade the implementation of the mathematics curriculum. Unlike heuristics and thinking skills, which have been a part of the framework for almost two decades now, reasoning, communications and connections, and applications and modelling have been rather "new" in the proposed tool kits of teachers and for many a challenge in some ways. In 2009, the Association of Mathematics Educators and the Mathematics and Mathematics Education Academic Group of the National Institute of Education organised a **Mathematics Teachers Conference** with the theme *Mathematical Applications and Modelling*. During the conference, several international and local mathematics educators shared with participants, mainly mathematics teachers, their understandings, knowledge and pedagogies related to mathematical applications and modelling. This book showcases many of the peer-reviewed and presented papers. The chapters of this book are an invaluable resource for teachers and educators in Singapore and elsewhere.

3 Mathematical Applications and Modelling

The syllabus documents of 2006 (Ministry of Education, 2006a; 2006b) for both primary and secondary schools, exemplify the process: *applications and modelling* as follows:

> Application and Modelling play a vital role in the development of mathematical understanding and competencies. It is important that students apply mathematical problem-solving skills and reasoning skills to tackle a variety of problems, including real-world problems.
>
> Mathematical modelling is the process of formulating and improving a mathematical model to represent and solve real-world problems. Through mathematical modelling, students learn to use a variety of representations of data and to select and apply appropriate mathematical methods and tools in solving real-world problems. The opportunity to deal with empirical data and use mathematical tools

for data analysis should be part of learning at all levels. (Ministry of Education, 2006a, p. 8; 2006b, p. 4)

The above does not appear to make any distinction between application and modelling but emphasise "solving real-world problems". **Stillman** in chapter 16 of this book makes an excellent and meaningful distinction between application and modelling. She states that

> In mathematical applications the task setter starts with mathematics and reaches out to reality. A teacher designing such a task is effectively asking: Where can I use this particular piece of mathematical knowledge? This leads to tasks that illustrate the use of particular mathematics content. They are a useful bridge into modelling but are not modelling in themselves. (p. 305)

> With mathematical modelling on the other hand, the task setter starts with reality and looks to mathematics before finally returning to reality to judge the usefulness and desirability of the mathematical model for description or analysis of a real situation. (p. 306)

From the syllabus documents it is apparent that the emphasis on applications and modelling is for all levels of schooling. Hence, the nature of applications and modelling tasks must be related to the mathematical knowledge of the group of students for whom the tasks are intended.

3.1 *Primary level*

In the primary school, it is heartening to note that students solve fairly difficult mathematical tasks as problem solving is the goal of the curriculum. Generally, the more challenging non-routine problems at this level can be considered as application problems. The intent of such tasks is to develop higher order thinking skills among the students. Textbooks, used in Singapore primary schools, contain both routine problems for exercises and non-routine problems for students to apply their knowledge of mathematics in new situations. However, in the textbooks, the non-routine problems are not labelled as application problems.

Since the advent of the mathematics curriculum for the New Education System in the 1990s, there has been great emphasis on thinking skills and problem solving heuristics, shown in Table 2.

Table 2
Thinking skills and heuristics for problem solving

Thinking Skills	Problem-Solving Heuristics
Classifying	Act it out
Comparing	Use a diagram/model
Sequencing	Make a systematic list
Analysing parts and whole	Look for pattern(s)
Identifying patterns and relationships	Work backwards
Induction	Use before-after concept
Deduction	Use guess and check
Generalising	Make suppositions
Verifying	Restate the problem in another way
Spatial Visualisation	Simplify the problem
	Solve part of the problem
	Thinking of a related problem
	Use equations

An example of a problem-solving task, students in the primary school may do is as follows:

The handshake problem
At a party there were 10 people. If everyone at the party shakes hands with everyone else, how many handshakes would there be?

There are various heuristics that can be used to solve this problem. Figure 2 shows the solution of a primary school student who obtained the solution by modelling the process, on paper in the form of a diagram, of how the handshakes were made. The problem solver has in a way executed the heuristic "act it out" on paper (see Kaur & Yeap, 2009a, p. 311).

Given that the focus of the curriculum has been on the processes of heuristics and thinking skills for the past two decades, teachers are more comfortable with solving of problems using problem-solving heuristics and thinking skills than applications and modelling. Teachers have received extensive training on the use of problem-solving heuristics, so much so, that the use of heuristics has been routinised through some exemplar problems. Students have thus developed some standard problem-solving procedures which are counter to the very idea of developing problem-solving skills. Also, textbooks and other assessment books provide a wide range of problems for students to practice and for teachers to adapt and use for their instructional needs.

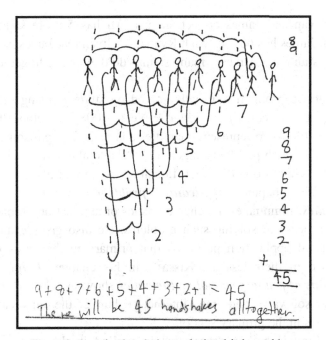

Figure 2. A student's solution to the handshake problem

As claimed by Reusser and Stebler (1997), at the primary level, students get to apply the mathematics that they have learned mostly through the so-called *word problems* or *story problems*. It is no different in the primary schools in Singapore. Students are exposed to word problems very early. An example of one such word problem is:

Tom has $18. Jenny has twice as much. How much money do they have altogether?

This word problem describes a seemingly real context for the child who has to understand the context and use appropriate mathematical operations to solve the problem. Such problems are common at the primary level in the curriculum.

More challenging word problems at the primary level are known as the infamous *Model Method* problems. An example of one such problem is as follows:

Ali, Ryan and James collect stamps. Ali has 5 more stamps than Ryan. James has 60% of what Ryan has. Given that James has half as many stamps as Ali, how many stamps do they have altogether?

The above problem is usually solved by students using the *Model Method* (see Ferrucci, Kaur, Carter & Yeap, 2008), a method that uses rectangular blocks to represent quantities and make comparisons. For the solution of all such problems, algebra is a viable alternative but students in primary school do not have sufficient know-how of algebra to do so. **Yeo** in his chapter: ***Application problems in primary school mathematics***, illuminates the characteristics of application problems and thinking process of solving such problems. He also gives examples of four types of application problems that primary mathematics teachers may infuse in their lessons. **Kissane** in his chapter: ***Using ICT in applications of primary school mathematics*** showcases how calculators, computer software and the internet enhance the application of mathematics in the primary school.

Modelling is not an aspect that is explicit in the primary school curriculum in Singapore. However, students are exposed to various types of models. Van de Walle (2004) claimed that a model for a mathematical concept refers to any object, picture, or drawing that represents the concept or onto which the relationship for that concept can be imposed. The Singapore mathematics curriculum emphasises that teachers use various types of models to illustrate mathematical concepts. For example, base ten blocks for numbers or area models for fractions are quite commonly used. The textbook series at the primary level however, do not have other types of modelling tasks and consequently the implementation of modelling in the classroom is left to individual teachers. **Dindyal** in his chapter: *Word problems and modelling in the primary school mathematics* explores the use of the modelling cycle in the solution process of word problems and classifies word problems from a modelling perspective. **Chan** in his chapter: *Mathematical modelling in a PBL setting for pupils: Features and task design* presents mathematical modelling as a problem-solving activity. He focuses on PBL as an instructional approach and the design of modelling tasks for such settings.

From the findings of TIMSS 2007 (Mullis, Martin & Foy, 2008) it is apparent that grade 4 (primary 4) teachers in Singapore spend their instructional time each week in the following ways. They have their students solve problems (35%), listen to lecture-style explanations (19%), review homework (14%), and take tests and quizzes (8%). They also re-teach content and procedures that students have difficulty with (11%), carry out classroom management tasks (6%), and other student activities (5%). There is little doubt that most, if not all of the problems students solve are of the "word problem" application type. On the contrary, at most 5% of the instructional time may have been used for some modelling activities. Modelling activities would require classroom discourse and organisation of student groups that are orthogonal to independent work or listening to lecture-style explanations.

Anthony and **Hunter** in their chapter: *Communities of mathematical inquiry to support engagement in rich tasks* draw on international and New Zealand classroom research. They claim when using modelling and application tasks that involve collaborative activity it is important that teachers ensure that group and classroom discussions support reasoned participation by all students. **Anderson** in her chapter: *Collaborative problem solving as modelling in the primary years of schooling* draws some similarities and differences between problem solving and modelling. She draws on appropriate examples and establishes that collaborative problem solving which requires the use of processes such as questioning, analysing, reasoning and evaluating to solve particular tasks mirrors mathematical modelling. Teachers themselves must experience mathematical modelling so as to understand the needs of their students engaged in mathematical modelling. **Ng** in her chapter: ***Initial experiences of primary school teachers with mathematical modelling*** suggests that more scaffolding was necessary to ease teachers into the implementation of such tasks in the primary classrooms, particularly in nurturing mindset change of teachers towards accepting multiple representations of a problem, diverse solutions, and what constitutes as mathematical in the representations.

3.2 Secondary level

In the secondary school, students are exposed to a wider range of mathematical topics, including some harder arithmetic topics. They gain more expertise in the use of algebra and geometry. Every topic they are taught culminates with application type of problems. Generally three types of application problems can be identified. The first type involves the application of mathematics from one content domain to another within mathematics. For example, the student who factorises 391 by writing it as 400 – 9 and using difference of two squares for factorisation is demonstrating this type of application. The second type can be considered as the application of mathematics in situations described as real-life but otherwise as purely mathematical problems. Most of the word problems used in mathematics falls in this category. The third type is problems in authentic situations in which the solver uses the mathematics that he or she knows to solve the problem in a real context. For example, a student who is in a store and has to decide which one is a better buy: a 400 g packet of biscuit or a 300 g packet of biscuit in a grocery store, given the price of each type of packet is solving this third type of application problem.

Students should demonstrate flexibility in solving all three types of application problems. Most of the so-called application problems in mathematics are of the first or second type. It is expected that students through solving the first and second type of application problems would be better prepared for solving the authentic problems (third type) that they would meet in their own daily lives. **YEO** in his chapter: ***Why study mathematics? Applications of mathematics in our daily life*** shares with readers several applications of mathematical knowledge and processes both in the workplace and outside the workplace. Examples in this chapter are helpful for teachers to bridge the knowledge their students construct during mathematics lessons and use of the knowledge in daily-life. Information and communication technology (ICT) provides tools for both teaching and learning in all schools in Singapore. **Kissane** in his chapter: ***Using ICT in applications of secondary school mathematics*** showcases how calculators, computer software and the internet enhance the application of mathematics in the secondary school. In addition, he also gives examples of how ICT tools may be used for simulations and modelling.

Just like their counterparts in the primary school, teachers in secondary schools are also more comfortable with problem-solving heuristics and thinking skills than applications and modelling. They too have received extensive training on the use of problem-solving heuristics, so much so, that the use of heuristics has been routinised through some exemplar problems. Students have also developed some standard problem-solving procedures which are counter to the very idea of developing problem-solving skills. Also, textbooks and other assessment books provide a wide range of problems for students to practice and for teachers to adapt and use for their instructional needs. Just like the primary level, modelling is not explicitly emphasised in the secondary mathematics curriculum. Textbooks do not address the process and interpretation of what constitutes modelling at the secondary level is left to individual interpretation of teachers. However, it would be inexact to say that students do not know about models. Models are used in the teaching of mathematics but modelling tasks are not explicitly used.

It may be said that the chapter in this book: ***Mathematical modelling in the Singapore secondary school mathematics curriculum*** by three curriculum specialists from the Ministry of Education, **Balakrishnan,**

Yen and **Goh**, is a noteworthy attempt to help secondary school mathematics teachers understand the nature of the mathematical modelling process and familiarise them with the four elements of the modelling cycle: mathematisation, working with mathematics, interpretation and reflection. **Kaiser**, **Lederich** and **Rau** in their chapter: *Theoretical approaches and examples for modelling in mathematics education* enlighten the reader with international perspectives and goals for teaching applications and modelling. They also give examples of educationally oriented modelling tasks and authentic complex modelling tasks. **Toh** in his chapter: *Making decisions with mathematics* proposes the use of decision-making activities to involve secondary school students in authentic tasks involving mathematical modelling. He demonstrates how this is possible through several examples.

From the findings of TIMSS 2007 (Mullis, Martin & Foy, 2008) it is apparent that grade 8 (secondary 2) teachers in Singapore spend their instructional time each week in the following ways. They have their students solve problems (34%), listen to lecture-style explanations (27%), review homework (11%), and take tests and quizzes (8%). They also re-teach content and procedures that students have difficulty with (9%), carry out classroom management tasks (6%), and other student activities (4%). There is little doubt that most of the problems students solve are typical textbook kind of application problems. On the contrary, at most 4% of the instructional time may have been used for some modelling activities. Modelling activities would require classroom discourse and organisation of student groups that are orthogonal to independent work or listening to lecture-style explanations.

Förster and **Kaiser** in their chapter: *The cable drum – Description of a challenging mathematical modelling example and a few experiences* detail how a modelling task was implemented in an upper secondary level class. This chapter sheds light on several aspects of classroom discourse and organisation that supports mathematical modelling. **Stillman** in her chapter: *Implementing applications and modelling in secondary school: Issues for teaching and learning* shares with readers her experiences from the perspective of teachers about the issues that relate to the implementation of applications and modelling and impact teaching and learning. She draws on her work in Australian schools and provides readers with empirical data to support her claims. Lastly, **Kemp** in her chapter: ***Developing pupils' analysis and interpretation of graphs and tables using a Five Step Framework*** provides teachers with a framework that helps students work with real-world problems and a variety of representations of data, particularly graphs and tables that are present in both print and digital media.

4 Conclusion

Teachers play an important role in implementing the curriculum. How can teachers be prepared to translate the goals of the curriculum into their lessons? Every revision of the mathematics curriculum brings about changes for teachers to address in their classrooms. In 1990, when the framework for the mathematics curriculum was introduced and emphasis placed on the use of heuristics and thinking skills as processes to engage students in mathematical problem solving, teachers attended briefings and workshops to equip themselves with the know-how. In tandem, pre-service courses for prospective mathematics teachers also embraced the framework of the mathematics curriculum and introduced teachers to the necessary processes. Textbooks also underwent revision. They too placed more emphasis on problems, problem-solving heuristics and thinking skills. Finally, a few years later assessment tasks in high stake examinations were also more problem solving oriented. So, it is not surprising that two decades after mathematical problem solving was made the goal of the school mathematics curriculum that teachers are comfortable with problems, problem-solving heuristics and thinking skills.

It is also worthy to note that the most common approach to teach problem solving adopted by teachers in Singapore is similar to what has been described by Lesh and Zawojewski (2007) about learning the content then the problem-solving strategies and then applying these. This approach has been described elsewhere as teaching for and about problem solving. Teaching via problem solving is not a popular local approach for teaching mathematics. As such, the local approach has evolved to what Stacey (2005) noted as

> ... promote good learning of routine content and to develop useful strategic and metacognitive skills, rather than explicitly to strengthen students' ability to tackle unfamiliar problems. (p. 349)

When the 2006 revised curriculum was introduced and emphasis placed on a process: reasoning, communication and connections Kaur and Yeap (2010) from the National Institute of Education carried out a professional development project: Enhancing the pedagogy of mathematics teachers to promote reasoning and communication in their classrooms (EPMT), over a two year period 2006 - 2008. The deliverables of the project, Kaur and Yeap (2009b, 2009c), and Yeap and Kaur (2010) have impacted the teaching and learning of mathematics in almost all schools in Singapore. With similar intent but in a different way, it is hoped that this book will contribute towards the development of teachers to embrace the process: applications and modelling and expand their repertoire of means to nurture mathematical problem solvers.

References

Ferrucci, B., Kaur, B., Carter, J. & Yeap, B. H. (2008). Using a model approach to enhance algebraic thinking in the elementary school mathematics classroom. In Greenes, C. E. & Rubenstein, R. (eds.), *Algebra and algebraic thinking in school mathematics: Seventieth yearbook* (pp. 195-210). Reston, VA: National Council of Teachers of Mathematics.

Kaur, B. & Yeap, B. H. (2009a). Mathematical problem solving in the secondary classroom. In P. Y. Lee & N. H. Lee (Eds.), *Teaching secondary school mathematics – A resource book* (pp. 305-335). McGraw-Hill Education (Asia).

Kaur, B. & Yeap, B. H. (2009b). *Pathways to reasoning and communication in the primary mathematics classroom*. Singapore: National Institute of Education.

Kaur, B. & Yeap, B. H. (2009c). *Pathways to reasoning and communication in the secondary mathematics classroom*. Singapore: National Institute of Education.

Kaur, B. & Yeap, B. H. (2010). *Enhancing the pedagogy of mathematics teachers project*. Singapore: National Institute of Education.

Lesh, R. & Zawojewski, J. (2007). Problem solving and modeling. In F. Lester (Ed.), *The second handbook of research on mathematics teaching and learning* (pp. 763-804). Charlotte, NC: Information Age Publishing.

Ministry of Education. (2006a). *Mathematics Syllabus: Primary*. Singapore: Author.

Ministry of Education. (2006b). *Mathematics Syllabus: Secondary*. Singapore: Author.

Mullis, I. V. S., Martin, M. O. & Foy, P. (2008). *International mathematics report: Findings from IES's Trends in International Mathematics and Science Study at fourth and eighth grades*. Boston, MA: Boston College.

Reusser, K. & Stebler, R. (1997). Every word problem has a solution – The social rationality of mathematical modelling in schools. *Learning and Instruction, 7(4)*, 309-327.

Stacey, K. (2005). The place of problem solving in contemporary mathematics curriculum documents. *Journal of Mathematical Behavior, 24*, 341-350.

Van de Walle, J. A. (2004). *Elementary and middle school mathematics: Teaching developmentally* (5th ed.). Boston: Pearson.

Yeap, B. H. & Kaur, B. (2010). *Pedagogy for engaged mathematics learning*. Singapore: National Institute of Education.

Part II
Applications & Modelling in Primary School

Chapter 2

Communities of Mathematical Inquiry to Support Engagement in Rich Tasks

Glenda ANTHONY   Roberta HUNTER

Providing students with rich learning tasks is one part of the picture. These rich tasks need to be enacted in ways that optimize the opportunities for diverse learners to develop and use a range of mathematical practices and understandings. When using modelling and application tasks that involve collaborative activity it is important that teachers ensure that group and classroom discussions support reasoned participation by all students. In this chapter we look at ways that teachers can model and orchestrate productive talk associated with rich tasks. We focus both on the participatory practices involved in a community of inquiry and on ways that teachers can support students in engage in mathematical argumentation.

1 Introduction

The creation of productive learning communities and the provision of rich tasks are two central elements of effective mathematics teaching (Anthony & Walshaw, 2007). Opportunities to learn mathematics depend significantly on both the community that is developed and on what is made available to learners. For all students the 'what' that they do in the classroom is integral to their learning. As Doyle (1983) notes, mathematical tasks draw students' attention towards particular mathematical concepts and provide information surrounding those concepts. It is by engaging with tasks that students develop ideas about the nature of mathematics and discover that they have the capacity to make sense of mathematics (Hiebert et al., 1997; Pepin, 2009).

Opportunities to work with rich tasks are important for students. Application and modelling tasks, in particular, require students to interpret a context and to make sense of the embedded mathematics (Stillman et al., 2009; Sullivan, Mousley, & Jorgensen, 2009). Task activity involves the application of a variety of mathematical practices such as testing conjectures, posing problems, looking for patterns, and exploring alternative solution paths. Moreover, because application and modelling tasks are typically posed in group work settings they also involve a range of metacognitive and communication practices. For example, Lesh and Lehrer (2003) describe the seventh-grade students' activities in solving *The Paper Airplane Problem* as follows:

> Sorting out and integrating concepts associated with a variety of different topic areas in mathematics and the sciences....they often emphasized multimedia representational fluency, as well as a variety of mathematical abilities related to argumentation, description, and communication—as well as abilities needed to plan, monitor, and assess progress while working in teams of diverse specialists. (p. 114)

In this chapter we draw on international and New Zealand classroom research (e.g., Hunter, 2007; Hunter & Anthony, in press) concerning productive mathematical communities that support student engagement in rich tasks. In our classroom studies the teachers used a communication and participation framework (see Appendix 1) as a flexible and adaptive tool to map out their development of an inquiry environment. The aim was for the teachers to support their students to develop progressively more proficient mathematical inquiry practices. In using the framework the teachers all reached similar endpoints at the conclusion of the study, but the individual pathway they each mapped out and traversed was unique (see Hunter, 2008). In this chapter, we focus on how the teacher can support students to develop effective ways of participating in mathematical discourse of inquiry—looking at ways of participating and communicating in group situations.

2 Learning Communities

Learning mathematics involves activity within a community. This claim falls naturally from theoretical framings that take as their key tenet the position that knowledge is socially constructed. The notion of a close relationship between social processes and conceptual development is fundamental to Lave and Wenger's (1991) social practice theory, in which the ideas of "a community of practice' and "the connectedness of knowledge" are central features. For Wenger (1998), participation in a community of practice is a "complex process that combines doing, talking, thinking, feeling, and belonging" (p. 56).

In accord with a focus on learning communities, recent curricula reforms (e.g., Kaur & Yeap, 2009; Ministry of Education, 2007; National Council of Teachers of Mathematics, 2000) advocate participation in mathematical discourse as a critical component of mathematical learning experience and practice. Carpenter, Franke, and Levi (2003) claim that learning mathematics entails "how to generate...ideas, how to express them using words and symbols, and how to justify to oneself and to others that those ideas are true" (p. 1). In advocating the role of equitable and productive learning communities Boaler's (2008) classroom research found that discussion in a group (or whole classroom setting) provides students with opportunities to share ideas and clarify understandings, develop convincing arguments regarding why and how things work, develop a language for expressing mathematical ideas, and importantly, learn to see things from other perspectives. Collaboration allows individual and collective knowledge to emerge and evolve within the dynamics of the spaces students and their teacher share; it supports the development of connections between mathematical ideas and between different representations of mathematical ideas (Cobb, Boufi, McClain, & Whitenack, 1997; Hunter, 2007; Sullivan et al., 2009; Watson, 2009). Of central interest to this chapter is the findings of Smith, Hughes, Engle, and Stein (2009) that "discussions that focus on cognitively challenging mathematical tasks, namely those that promote thinking, reasoning, and problem solving, are a primary mechanism for promoting conceptual understanding of mathematics" (p. 549).

Whilst the provision of mathematics classroom tasks that require high-level cognitive thinking is a hallmark of effective teaching (Anthony & Walshaw, 2009), it is only the first part of the learning scenario. Productive engagement in mathematical discourse associated with rich tasks can only happen in classrooms that have established appropriate inclusive participatory practices and general and mathematical obligations (Cobb, Gresalfi, & Hodge, 2009, Walshaw & Anthony, 2008). Stein, Grover, and Henningsen (1996) offer a conceptual framework, (see Figure 1) which illustrates the relationship among various task-related variable and students' learning outcomes.

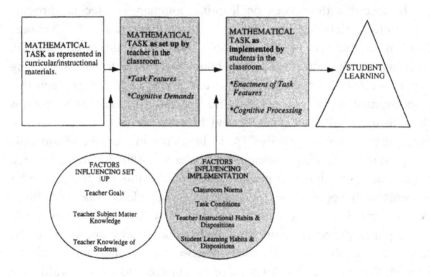

Figure 1. Adapted from Stein et al.'s (1996) Task implementation framework

Although the variables affecting task enactment are numerous, and each in their own way significant, in this chapter we focus on the shaded area with respect to establishment of classroom norms and teacher and student learning dispositions. First we discuss how teachers can create effective learning communities that encourage participation in inquiry discourse, and then look more closely at how teachers can support students to develop effective mathematical argumentation practices—including making mathematical conjectures, explanations, generalisations, and justifications.

3 Organising Learning Communities

Learning mathematics within classrooms is an inherently social process. Within a group task discussion provides students with a vehicle to extend their understanding of ideas or solutions methods. The "interaction of a variety of alternative conceptual systems that are potentially relevant to the interpretation of a given situation" (Lesh & Zawojewski, 2007, p. 789) is an essential mechanism for moving learning beyond current ways of thinking. However, the productiveness of any group is very much dependent on the ability of individuals to function within a group activity. Collaborative group work demands a range of social and cognitive learning behaviors; students must pose numerous questions and conjectures, engage in conflict resolution, and revise their thinking. Moreover, when the product of a task is required to be shared with and used by others—as is frequently the case in modelling tasks—the group must assess and prepare to communicate their products (English, 2006). Students need to understand that they must take responsibility for contributing, active listening, and sense-making (Doerr, 2006).

To illustrate the teacher's role in organising and supporting students to participate in communication practices associated with collective inquiry we draw on our current research with teachers in primary schools in New Zealand (e.g., Hunter, 2007; Hunter & Anthony, in press) and international research studies. Our focus is the role of the teacher in negotiating norms and obligations regarding participation, risk taking and positioning of student contributions.

3.1 *Clarifying participations rights and obligations*

The responsibility to establish a supportive and productive learning community is, in the first instance, the teacher's. For this to happen, students need a safe, supportive learning environment that promotes social and intellectual risk-taking. In attending to students' 'mathematical relationships' (Black, Mendick, & Solomon, 2009) teachers and students should engage in explicit discussion about their participation rights and obligations concerning contributing, listening, and valuing; that is, there needs to be a common understanding of what it means to be accountable to the learning community.

First and foremost is the requirement to establish conditions for respectful discourse. Discourse is respectful when each person's ideas—be it student or teacher's—are taken seriously. Respectful discourse is also inclusive; all contributions are valued and no one person is disregarded. To create such an environment, teachers need to ensure that everyone is willing to contribute and that others will listen carefully (Chapin & O'Connor, 2007). In Hunter's (2007) study of primary level classes she found that expectations were most effective when established through negotiation with students. Examples of negotiated rights and responsibilities included: the right of all students to contribute and to be listened to; the right to test out ideas that may or may not be correct without fear of having other students making disrespectful comments; and the right to have other people discuss your ideas and not you. To affirm these rights the teachers implemented specific strategies to re-mediate unproductive situations in their existing classroom culture. They took care in the first instance to engineer learning partnerships (e.g., placing students in supportive groupings), and reinforced the use of talk–formats that valued assertive communication ("We ask for reasons why"), encouraged construction of multiple perspectives ("We respect other people's ideas and do not just use our own"), and frequently affirmed valuing individual contribution ("We think about all the different ways before a decision is made about the group's strategy solution"). Disruptive and disparaging behavior in group work, or put downs in whole class reporting was quickly established as not acceptable behavior.

3.2 *Supporting students to take risks*

In rich tasks, where many different ideas draw on a range of mathematical concepts, in-depth discussion provides students with opportunities to extend their understanding of ideas or solutions methods.

However, having to share ideas publically involves taking risks. Open-ended activities and modelling type activities, in particular, involve putting ideas out there, tossing ideas around, making conjectures, and sometimes going down the wrong track. When students present incorrect ideas based on faulty reasoning or misconceptions it is possible that they feel vulnerable and may see another student's disagreement with their ideas as a judgment about themselves. For instance, Chapin and O'Connor (2007) remind us that a retort such as, "I don't think Jasmine's method will work" may, for some students, be equated with "I don't think Jasmine ever has good ideas" (p. 125). Teachers need, therefore, to be vigilant of the potential social consequences of participating in group activities in this way. Changing the focus of attention to the 'idea' that has been presented rather than the person is one way to assist students to evaluate contributions according to their mathematical validity or accuracy and reinforce expectations of collective sense-making and justification.

A teacher in our research talked explicitly to her students about risk taking using an analogy of being in one's comfort zone:

Remember how yesterday we talked about in maths learning how you go almost to the edge? So I'm going to move you out of your comfort zone...out a little bit more, and then a little bit more. And when you are out there you will make that your comfort zone.

In using the comfort zone analogy the teacher suggested that the process of opening up one's thinking for inspection would be difficult at first, but with more time it would become a natural and valued part of doing and learning mathematics.

3.3 *Supporting students to be positioned competently*

We have seen that rich tasks provide many ways for students to contribute ideas. The multidimensional nature of mathematical work involved in modelling and applications problems include: making connections across ideas; rephrasing and re-presenting problems; finding patterns; using models and manipulatives; asking questions and making conjectures, offering knowledge of real situations; offering alternative or partial solutions; and determining the efficiency of solutions. To allow and encourage students to engage in 'mathematical play' (Holton, Ahmed, Williams, & Hill, 2001) every student needs to feel that she or he is somebody with good ideas.

In addition to reaffirming the participation rights and obligations of group activity, the teacher can take a proactive role in highlighting contributions from less able or less vocal students, making sure that their thinking is positioned as valuable. For example, when a teacher in our study overheard a comment by a low achieving student she commented:

> *Wow, Teremoana see how you have made them think when you said that? Now they are using your thinking.*

Positioning students as someone with good ideas in this broader sense, disrupts those traditionally narrow ways of being competent in mathematics like finishing first or being born with mathematical ability (Kazemi & Hintz, 2008). Importantly, studies promoting relational equity (e.g., Boaler, 2008) also stress the value of appreciating the diversity in students' contributions, different ways of thinking, and different viewpoints.

The teacher can also take care to ensure that academically productive talk is not just for those students with strong verbal skills or who are confident about speaking out. In some groups there are individuals or groups of students who might initially prefer to remain passive. For example, in New Zealand schools Pasifika girls may be reluctant to speak up or to question a boy's thinking. Without affirmative support they are likely to be more comfortable in the role of listening respectfully to the teacher. In our research classrooms, an example of proactive action taken by a teacher to a Pasifika girl's contribution was:

> *You don't have to whisper. You can talk because we want to make sure that you are heard.*

On another occasion a student was told to "speak up, I like the way you are thinking but we need to hear you". By regularly calling on students to respond to the ideas are being discussed, regardless of whether they volunteered, the teacher reaffirms that all students are expected to make sense of these problems and all students are expected to participate.

4 Facilitating Mathematical Discourse

Rich tasks provide opportunities for students to construct mathematical understandings in conjunction with developing skills in mathematical argumentation. Specifically, group engagement requires students to develop sharable products that involve descriptions, explanations, justification, and mathematical representations. In learning to explain and justify their reasoning students' conceptual explanations form an important precursor for developing explanatory justification and argument (Cobb et al., 1997). To assist student to develop well-structured explanations it is important for teachers to provide models of good explanations—explanations that have a conceptual, not calculational basis, with reasoning that is clear, visible, and available for question, clarification, or challenge by others. Teachers can also utilize a range of 'talk moves' (O'Connor, 2001) to highlight productive and explanatory talk. These include: revoicing, eliciting students' reasoning, and modelling mathematical language. Importantly, we have found that when the teachers use these practices students quickly adopt these same practices in their interactions within group activities, and with more experience, within whole-class report back sessions.

4.1 *Revoicing*

Revoicing—repeating, sometimes in a re-phrased format is an action that can be used for multiple purposes. It may be used to clarify a student's meaning, to focus the attention of others on an important mathematical idea, to help a student clarify their thinking, or as an opportunity to extend a student's mathematical thinking. For example, Ava a teacher in Hunter's (2007) study uses revoicing as follows:

Ava: Rachael was saying she is adding three, adding another three, so that's three plus three plus three. So if you keep adding three all the time what is another way of doing it?

Alan: You can just times instead of adding. It won't take as long and it is more efficient.

Ava: Yes, you are right. Did you all hear that? Alan said that you can just times it, multiply by three because that is the same as adding on three each time. What word do we use instead of timesing?

Alan: Multiplication, multiplying.

Here we see that the teacher accepted the students' use of colloquial terms but revoiced using rich multi-levels of mathematical language.

4.2 *Eliciting students' reasoning*

When working in groups, students need to ask mathematical questions of one another. They first learn to do this by interacting in discussions with the teacher. Teachers can model and support student use of questions which clarify or extend aspects of an explanation with questions starters like: What did you do there…? Where did you get that number from? Can you show, draw, or use materials to illustrate what you did? Questions that probe students to offer justifications include: Why did you…? So what happens if…? But how do you know it works? Can you convince us? So why is it that…? Other times, questions can be used to support students to make connections between mathematical ideas and test generalisations; for example: Does that always work? Can you give us a similar example? Is it always true? Can you link all the ideas you have used?

The following episode illustrates how students adopt question prompts in their group activities to develop shared understanding of all of the group members:

Aroha: *[records 43, 23, 13, 3 and then 3 x 4 = 12] I am adding forty-three, twenty-three, thirteen, and three, so three time fours equals twelve.*

Kea: *What are you trying to do with those numbers? Where did you get the four?*

Donald: *All she is doing is like making it shorter by like doing four times three.*

Hone: *Because there are only the tens left.*

Donald: *Three times four equal twelve and she got that off all the threes, like the forty-three, twenty-three, thirteen, and three. So she is just like adding the threes all up and that equals twelve.*

Knowing when and how to step into a group discussion can be difficult for teachers—as traditionally teachers have been the ones to correct thinking or provide answers (Lobato, Clarke, & Ellis, 2005). In the following example the teacher intercedes in a group discussion by modelling questions and eliciting students to provide conceptual rather than procedural explanations. In solving the problem of sharing three cakes between eight people the teacher enters the group discussion as follows:

Hone: *What are you doing?*

Anaru: *Twenty four eighths.*

Teacher: *But I am not sure...we know what you mean. Can you explain it?*

Anaru: *[Frowned and shook her head in response]*

Hemi: *[points at the symbols 24/8 which Anaru had recorded next to the drawing] I can. Twenty four eighths, because there are eight in each cake and there's eight slices in each cake and it all adds up to twenty four.*

Teacher: *Twenty four what? What does that bottom number mean?*

Heni: *That means how much slices in each cake.*

Teacher: *Okay, what does twenty four represent?*

Hemi: *It means how much altogether.*

Teacher: *Altogether, Yeah, twenty four bits, slices and they are all eighths.*

Here we see how the teacher focused on the mathematical meaning by directing students' attention to very specific aspects of their explanation [what does the twenty four mean?], requiring the students to relate their explanation to the problem at hand. Additionally, the teacher, working with these students early in the school year, reaffirmed a supportive environment, one in which peers were expected to be able to offer support in helping clarify thinking.

When involving students in responding to teachers' questions, or questioning others, it is important that students are held accountable to listen and make sense of the mathematical explanations. One way that teachers (and students in their groups) can do this is by asking students if they agree or disagree with the results or ideas and why. Another way is to constantly reaffirm the norms of participation, inclusive of the expectation of mathematical sense-making. The following episode illustrates how a teacher reminded her students of their role in sense-making:

> [In your groups you need to]...*talk about what you are doing...so whatever numbers you have chosen don't just write them. You say, I am going to work with...or I have chosen this and this because...and this is what I am going to do...you need to explain how you are working it out to your group. They are going to listen. I want you to think about and explain what steps you are doing, what maths thinking you are using. The others in the group need to listen carefully and stop you and question any time or at any point where they can't track what you are saying.*

In expecting students to shape conjectures, respond to questions, or examine the thinking of others, it is also important that they are provided with 'think time' or as one teacher in our study labeled it 'rethink time'. For example, in seeing a student struggling to develop an explanation the teacher said to the group:

> *I can see you are confused. Me too, that's all right we can take some time to rethink about it. It's good to take some risks with our thinking sometimes.*

The teacher's validation of the acceptability of confusion as a natural part of sense-making normalized that learning mathematics involves effort and sometimes struggle.

5 Managing Whole-Class Discussions

In rich tasks there is frequently a requirement that groups report their collective thinking to the whole class. Here again the teacher plays a vital role in orchestrating the discussion. The teacher needs to guide students to draw connections among the methods and maintain a focus on the key disciplinary ideas inherent in task. Smith et al. (2009) note the importance of the teacher anticipating the likely responses of students to the task, including those which use erroneous thinking. Anticipation of student responses assists the teacher to monitor students' work on, and engagement with, the task while they are working in groups and informs the teacher's selection of particular students to present their mathematical work in the larger sharing session. Careful selection enables the teacher to match or extend the mathematical focus of the problem, add another dimension to, or provide opportunities for deeper exploration and analysis of students' thinking. Connecting different student responses and connecting the responses to key mathematical ideas helps the teacher to advance the mathematical understanding of the class as a whole. As in the small group situation, it is important that the teacher affirm the participatory obligations and provide spaces for questions, clarification, and explanatory justification.

While on one level, whole class reporting provides opportunities for the teacher and other students to assess learning progress, discussions about rich problems can sometimes advance into areas that have not been previously covered in the classroom curriculum. For example, in the following episode involving 4th graders discussion of a model for selecting a swim team for the Olympics, English (2006, p. 200) notes how the whole-class discussion highlighted and expanded on the group's emergent understanding of the 'mean':

Teacher: *Why were the averages important? And how did you work out averages?*

James: *We worked it out by adding up all the scores, adding up all the times, and then dividing it by how many events they had competed in.*

Teacher: *....you told me you divided the first by three because...*

Luke: *That's how many times the person actually swam.*

Teacher: *Did all swimmers do three races?*

Luke: *No.*

Teacher: *If somebody swam in four races what did you do then?*

Luke: *You had to divide it by four because it's not really fair if ... someone competes three times and the other person competes like eight times because eight times could get a higher amount.*

Here we see how the teacher's press for the group members to explain and justify their reasoning enabled their processes of thought as well as products to become visible.

6 Conclusion

We want all students involved in rich mathematical tasks to believe in themselves as mathematical thinkers, and to be able to build their knowledge, skill, and identities as successful mathematical students. To do so, requires that teachers and students develop shared expectations for participating in mathematical discussions and that teachers provide clear and explicit prompts and modelling of mathematical practices associated with argumentation. Mathematical modelling and application problems that are solved in group settings provide a solid basis for student development of argumentation because they are inherently social experiences—experiences that with effective teacher support and guidance can foster effective communication, teamwork, and reflection. Learning experiences involving students working through complex circumstances collaboratively as a member of a team are likely to develop skills for future-orientated learning in the 21st century.

References

Anthony, G., & Walshaw, M. (2007). *Effective pedagogy in mathematics/pāngarau: Best evidence synthesis iteration [BES]*. Wellington: Ministry of Education.

Anthony, G., & Walshaw, M. (2009). *Effective pedagogy in mathematics,* No 19 in the International Bureau of Education's Educational Practices Series: Available at www.ibe.unesco.org/en/services/publications/educational-practices.html.

Black, L., Mendick, H., & Solomon, Y. (Eds.). (2009). *Mathematical relationships in education: Identities and participation.* New York: Routledge.

Boaler, J. (2008). Promoting 'relational equity' and high mathematics achievement through an innovative mixed-ability approach. *British Educational Research Journal, 34*(2), 167-194.

Carpenter, T., Franke, M., & Levi, L. (2003). *Thinking mathematically: Integrating arithmetic and algebra in elementary school.* Portsmouth, NH: Heinemann.

Chapin, S. H., & O'Connor, C. (2007). Academically productive talk: Supporting students' learning in mathematics. In W. G. Martin, M. Strutchens, & P. Elliot (Eds.), *The learning of mathematics* (pp. 113-139). Reston, VA: NCTM.

Cobb, P., Boufi, A., McClain, K., & Whitenack, J. (1997). Reflective discourse and collective reflection. *Journal for Research in Mathematics Education, 28*(3), 258-277.

Cobb, P., Gresalfi, M., & Hodge, L. L. (2009). An interpretive scheme for analyzing the identities that students develop in mathematics classrooms. *Journal for Research in Mathematics Education 40*(1), 40-68.

Doerr, H. M. (2006). Teachers' ways of listening and responding to students' emerging mathematical models. *ZDM Mathematics Education, 38*(3), 255-268.

Doyle, W. (1983). Academic work. *Review of Educational Research, 53*, 159-199.

English, L. (2006). Introducing young children to complex systems through modelling. In P. Grootenboer, R. Zevenbergen, & M. Chinnappan (Eds.), *Identities cultures and learning spaces* (Proceedings of the 29th annual conference of the Mathematics Education Research Group of Australasia, Vol. 1, pp. 195-202). Sydney: MERGA.

Hiebert, J., Carpenter, T., Fennema, E., Fuson, K. C., Wearne, D., Murray, H., Olivier, A., & Human, P. (1997). *Making sense: Teaching and learning mathematics with understanding.* Portsmouth, NH: Heinemann.

Holton, D., Ahmed, A., Williams, H., & Hill, C. (2001). On the importance of mathematical play. *International Journal of Mathematical Education in Science and Technology, 32*(3), 401-415.

Hunter, R. (2007). *Teachers developing mathematical practices in communities of inquiry*. Unpublished Doctoral thesis, Massey University, Palmerston North, New Zealand.

Hunter, R. (2008). Facilitating communities of mathematical inquiry. In M. Goos, R. Brown, & R. Makar (Eds.), *Navigating currents and charting directions* (Proceedings of the 31st annual Mathematics Education Research Group of Australasia conference, Vol. 1, pp. 31-39). Sydney: MERGA.

Hunter, R., & Anthony, G. (in press). Developing mathematical inquiry and argumentation In R. Averill & R. Harvey (Eds.), *Teaching Primary School Mathematics and Statistics: Evidence Based practice*. Wellington: New Zealand Council of Educational Research.

Kaur, B., & Yeap, B. H. (2009). *Pathways to reasoning and communication in the primary school mathematics classroom*. Singapore: National Institute of Education.

Kazemi, E., & Hintz, A. (2008). *Fostering productive mathematical discussions in the classroom*. Manuscript submitted for publication.

Lave, J., & Wenger, E. (1991). *Situated learning: Legitimate peripheral participation*. Cambridge, UK: Cambridge University Press.

Lesh, R., & Lehrer, R. (2003). Models and modeling perspectives on the development of students and teachers. *Mathematical Thinking and Learning, 5*(2 & 3), 109-129.

Lesh, R., & Zawojewski, J. (2007). Problem solving and modeling. In F. K. Lester (Ed.), *Second handbook of research on mathematics teaching and learning* (pp. 763-804). Charlotte, NC: Information Age.

Lobato, J., Clarke, D., & Ellis, A. (2005). Initiating and eliciting in teaching: a reformation of telling. *Journal for Research in Mathematics Education, 36*(2), 101-136.

Ministry of Education (2007). *The New Zealand Curriculum*. Wellington: Learning Media, Ministry of Education.

National Council of Teachers of Mathematics (2000). *Principles and standards for school mathematics*. Reston, VA: National Council of Teachers of Mathematics.

O'Conner, M. (2001). "Can any fraction be turned into a decimal?" A case study of a mathematical group discussion. *Educational Studies in Mathematics, 46*(1-3), 143-185.

Pepin, B. (2009). The role of textbooks in the 'figured worlds' of English, French, and German classrooms. In L. Black, H. Mendick, & Y. Solomon (Eds.), *Mathematical relationships in education* (pp. 107-118). New York: Routledge.

Smith, M. S., Hughes, E. K., Engle, R. A., & Stein, M. K. (2009). Orchestrating discussions. *Mathematics Teaching in the Middle School, 14*(9), 549-556.

Stein, M., Grover, B., & Henningsen, M. (1996). Building student capacity for mathematical thinking and reasoning: an analysis of mathematical tasks used in reform classrooms. *American Educational Research Journal, 33*(2), 455-488.

Stillman, G., Cheung, K.-C., Mason, R., Sheffield, L., Sriraman, B., & Ueno, K. (2009). Challenging mathematics: Classroom practices. In E. J. Barbeau & P. J. Taylor

(Eds.), *Challenging mathematics in and beyond the classroom: The 16th ICMI study* (pp. 243-283). New York: Springer.

Sullivan, P., Mousley, J., & Jorgensen, R. (2009). Tasks and pedagogies that facilitate mathematical problem solving. In B. Kaur, B. H. Yeap, & M. Kapur (Eds.), *Mathematical problem solving* (pp. 17-42). Singapore: World Scientific.

Walshaw, M., & Anthony, G. (2008). The role of pedagogy in classroom discourse: A review of recent research into mathematics. *Review of Educational Research, 78*(3), 516-551.

Watson, A. (2009). Modelling, problem-solving and integrating concepts *Key understandings in mathematics learning*. Oxford: Nuffield Foundation.

Wenger, E. (1998). *Communities of practice: Learning, meaning and identity*. Cambridge, UK: Cambridge University Press.

Appendix 1:
Communication and Participation Framework

	Phase One	Phase Two	Phase Three
Making conceptual explanations	Use problem context to make explanation experientially real.	Provide alternative ways to explain solution strategies.	Revise, extend, or elaborate on sections of explanations.
Making explanatory justification	Indicate agreement or disagreement with an explanation.	Provide mathematical reasons for agreeing or disagreeing with solution strategy. Justify using other explanations.	Validate reasoning using own means. Resolve disagreement by discussing viability of various solution strategies.
Making generalisations	Look for patterns and connections. Compare and contrast own reasoning with that used by others.	Make comparisons and explain the differences and similarities between solution strategies. Explain number properties, relationships.	Analyse and make comparisons between explanations that are different, efficient, sophisticated. Provide further examples for number patterns, number relations and number properties.
Using representations	Discuss and use a range of representations to support explanations.	Describe inscriptions used, to explain and justify conceptually as actions on quantities, not manipulation of symbols.	Interpret inscriptions used by others and contrast with own. Translate across representations to clarify and justify reasoning.
Using mathematical language and definitions	Use mathematical words to describe actions.	Use correct mathematical terms. Ask questions to clarify terms and actions.	Use mathematical words to describe actions. Reword or re-explain mathematical terms and solution strategies. Use other

			examples to illustrate.
Participatory actions	Active listening and questioning for more information. Collaborative support and responsibility for reasoning of all group members. Discuss, interpret and reinterpret problems. Agree on the construction of one solution strategy that all members can explain. Indicate need to question during large group sharing. Use questions which clarify specific sections of explanations or gain more information about an explanation.	Prepare a group explanation and justification collaboratively. Prepare ways to re-explain or justify the selected group explanation. Provide support for group members when explaining and justifying to the large group or when responding to questions and challenges. Use wait-time as a think-time before answering or asking questions. Indicate need to question and challenge. Use questions which challenge an explanation mathematically and which draw justification. Ask clarifying questions if representation and inscriptions or mathematical terms are not clear	Indicate need to question during and after explanations. Ask a range of questions including those which draw justification and generalised models of problem situations, number patterns and properties. Work together collaboratively in small groups examining and exploring all group members reasoning. Compare and contrast and select most proficient (that all members can understand, explain and justify).

Chapter 3

Using ICT in Applications of Primary School Mathematics

Barry KISSANE

Apart from easing some of the computational burden, an attraction of ICT for mathematics education is that we may have new opportunities for pupils to learn and teachers to teach. This chapter describes examples of technologies likely to be appropriate for primary school use in Singapore, and analyses some of their potential for applications of mathematics in particular. Calculators are important because they are the most likely to be affordable and available, and have now been sanctioned for use in Singapore schools for Primary 5 and Primary 6. Computer software with particular strengths is acknowledged, including general purpose software and *Tinkerplots*, an innovative software package for statistics, aimed at pupils in the middle years. The Internet offers considerable potential for mathematics education and seems likely to continue to grow in significance in Singapore. Some ways in which the Internet might support increased attention to applications of mathematics are identified, including access to data analysis tools, simulation devices and examples of applications.

1 Introduction

There have been remarkable changes in the past decade to the Information and Communications Technology (ICT) available to pupils, teachers and schools and, indeed, to the wider Singaporean society. These tools include calculators and computers, together with their software and the Internet. Such changes together suggest that a 21st century mathematics curriculum should be substantially different from curricula of earlier times. While many uses of ICT are focused on 'teaching' pupils (for example, using computer software to provide systematic practice of skills), this chapter focuses on some of the ways in which pupils might use ICT in relation to applications of mathematics in particular. These include representing real-world situations, undertaking computation, collecting and analysing data, using simulation, and describing applications. Examples are provided of some of the ways in which ICT use by pupils can contribute to their understanding of and competence in applying mathematics to the real world, especially in probability and statistics.

Recent changes to the primary school syllabus for mathematics in Singapore, referred to subsequently in this chapter as Primary Syllabus, have included attention to ICT. For example, the Singapore Ministry of Education (2006, p. 5) has referred explicitly to the use of ICT and applications in describing one of eight aims of mathematics education in schools as to enable pupils to "[m]ake effective use of a variety of mathematical tools (including information and communication technology tools) in the learning and application of mathematics."

As well as having some intrinsic importance, applications play a vital role in the development of mathematical understanding and competencies, as reflected in the Syllabus. It is important that pupils apply mathematical problem-solving skills and reasoning skills to tackle a variety of problems, including real-world problems. Their learning of mathematics and their understanding of its relevance to modelling will be supported by using mathematical ideas and thinking. The Primary Syllabus acknowledges the significance of pupil use of calculators in particular:

> Calculators and other technology tools are tools for learning and doing mathematics. The introduction of calculators at P5 [Primary 5] and P6 [Primary 6] reflects a shift to give more focus to processes such as problem solving skills. The rationale for introducing calculators at the upper primary levels is to:

1. Achieve a better balance between the emphasis on computational skills and problem solving skills in teaching and learning and in assessment
2. Widen the repertoire of teaching and learning approaches to include investigations and problems in authentic situations
3. Help pupils, particularly those with difficulty learning mathematics, develop greater confidence in doing mathematics.

(Ministry of Education, 2006, p. 11)

It is also worth noting here that, despite the reservations of some teachers and others, credible research evidence regarding anticipated harmful effects of the use of calculators has not been found and, indeed, researchers have long ago stopped investigating this question. Thus, as far back as 1992, Kaput, in a major review surveying the field of technology in mathematics education, dismissed the entire issue in a single paragraph, the first half of which noted:

The Hembree and Dessart (1986) meta-analysis study of the impacts of calculator use and the study by Wynands (1984) of several thousand youngsters in Sweden are convincing, at least to researchers, that heavy use of calculators in the early grades as part of instruction and assessment does not harm computational ability and frequently enhances problem-solving skill and concept development (Kaput, 1992, p. 534).

Since Kaput's work, attention has also been paid to the use of calculators by pupils in the early years of school, with similar positive findings regarding their effects on pupil learning (Groves, 1997).

Although mathematical modelling is not often explicitly referred to in primary schools, applications of mathematics always assume a mathematical modelling process of some kind. A useful representation of the thinking involved is shown in Figure 1 (Burkhardt, 1981).

With this conception of modelling in mind, it is important to note that the use of ICT in applications of mathematics occurs only in two of Burkhardt's four phases of modelling: formulating a model within an

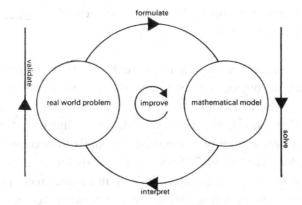

Figure 1. Mathematical modelling (from Burkhardt, 1981)

ICT context and using the ICT to 'solve' are the key features. The rest of the process of mathematical modelling involves problem-solving and thinking skills.

2 A Variety of ICT Tools

The Primary Syllabus (Ministry of Education, 2006) recognises the importance of a range of technologies for mathematical work:

- The introduction of calculators would not take away the importance of mental and manual computations. These skills are still emphasised as pupils need to have good number sense and estimation skills to check the reasonableness of answers obtained using the calculator.
- The use of scientific calculators is encouraged, although only the basic keys are required. Pupils can continue to use the same calculator when they are in secondary schools.
- There are many softwares and open tools that teachers can use to enhance the teaching and learning of mathematics. Basic word processing and spreadsheets skills would support the different strategies and activities that pupils could engage in. Technology also has an inherent motivating and empowering effect on pupils.

Teachers and pupils should harness technology in the teaching and learning of mathematics. (p. 11)

Although a variety of ICT tools might be used in applications of mathematics, an object is only likely to serve as a tool for pupils if attention is given to learning how to use it effectively, efficiently and autonomously. Indeed, the Primary Syllabus (Ministry of Education, 2006, p. 7) refers explicitly to the need for pupils to develop the skill to "use technology confidently". So some emphasis on the nature and role of the tool is needed in classrooms. Indeed, in general, mere provision of ICT is unlikely to be sufficient; it seems important for attention (and time) to be given to the development of expertise with appropriate ICTs as part of the tools of the mathematical trade for all pupils. The Singapore Mathematics Framework (Ministry of Education, 2006, p. 6) explicitly recognises the importance of skill development for the use of mathematics tools. Some comments about the range of possible tools is offered below.

2.1 *Arithmetic calculators*

Arithmetic computation can be handled by people using machines (such as a computer, a calculator, or an abacus) or by people not using machines (such as mental or written computation). Personal calculations can be completed mentally (based on understanding the structure of the number system and relations between numbers, as well as the recall of some number facts) or by hand on paper (which also involves mental work, of course, as well as learned efficient procedures) or with a machine (which also involves mental work, making choices about how best to use the device). All of these alternatives have a place in any primary curriculum, and it is important that pupils develop the confidence and competence to choose sensibly between these when applying mathematics. This includes learning to make decisions for themselves about when, where and why to use calculators.

An arithmetic calculator such as those used by many shops and traders in markets will allow pupils to undertake numerical computations when the tasks are beyond their mental capabilities, or when accurate results (rather than rough approximations) are needed efficiently. Since real-world data of the kinds expected in applications of mathematics will usually not be 'cleaned up', many computations encountered are best handled with a calculator. As well as when and why to use them, pupils need to learn how to use basic calculators, which are part of the everyday world of commerce, and hence part of using mathematics in everyday situations. Depending on the calculators, this is not usually a difficult matter, but requires attention to details, such as how calculators represent numbers (usually a little differently from handwritten or printed numbers), how negative numbers are represented, how (or whether) the rule of order is followed, how memories and percentage keys work (with the latter frequently very problematic) and how to deal with repeated operations. None of these ought be taken for granted, or assumed to be self-evident. Overall, however, the mathematical capabilities of these calculators are limited to arithmetic computation and hence unlikely to be sufficient for pupil needs beyond the early primary years.

2.2 *Scientific calculators*

A scientific calculator is easily recognisable as it usually allows for 'table' functions, such as trigonometric or logarithmic functions, to be readily evaluated, obviating the need for a book of mathematical tables. Such functions are rarely important for primary school pupils however. It will also extend the range of numbers that can be accommodated, to very large and very small numbers, because of the availability of scientific notation. They also permit more sophisticated mathematical operations such as raising numbers to powers and finding roots; recent models also allow for some elementary data analysis, usually restricted to finding sample statistics such as means and standard deviations. Calculators of this kind have been in widespread use in secondary school curricula in many countries for some thirty years or so and regarded as standard mathematical tools for all pupils. However, scientific calculators are essentially tools for calculation of numerical results and do not provide much more support than do arithmetic calculators for the purpose of applying mathematics.

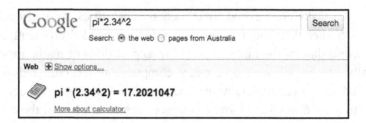

Figure 2. The use of *Google* as a scientific calculator

Like arithmetic calculators, scientific calculators are becoming very common in the everyday world, although rarely in shops and markets. Rather they are now almost standard equipment on mobile phones and even on the Internet. To illustrate the latter, Figure 2 shows the use of the search engine *Google* to find the area of the top of a large water tank with diameter measured as 4.68 m. As for hand-held calculators, the use of such facilities will require pupils to develop expertise to input information efficiently, choose suitable operations and interpret results correctly.

2.3 *Educational calculators for primary school*

Recently, calculators that are more powerful than arithmetic calculators, yet less powerful than scientific calculators have been developed especially for educational needs. A current model is the Texas Instruments TI-15 calculator, intended for use in the upper primary school. A typical capability of such calculators is a suite of ways of handling fractions as well as decimals; so they allow smooth transition between fractions, decimals and percentages, and facilitate both arithmetic operations with fractions and simplification of fractions. Kissane (1997) has argued that calculators of these kinds meet the needs of upper primary pupils better than either arithmetic or scientific calculators, as they are explicitly designed for that purpose.

The facility of a calculator to smoothly convert between fractions and decimals, allowing the process of 'simplifying' fractions to be carried out in a transparent way, is likely to be of considerable conceptual help to pupils in P5 and P6, helping them to recognise that fractions and decimal representations of numbers are merely representations, and not in fact different numbers (as some seem to think). While pupils are expected to calculate with fractions without a calculator, the process of learning this can be well supported by calculators. In addition, while standard algorithms for addition and subtraction can sometimes have the effects of pupils regarding a fraction as a pair of numbers (rather than as a single number), the use of these calculators may help them experience more productively the essential idea of equivalence of fractions, which underpins the standard algorithms. In these senses, the technology has moved beyond being merely of computational help, but also of importance for developing a rich conceptual understanding of the mathematical ideas involved.

2.4 Computer software

While computers tend to be larger, more expensive and less portable than calculators, they have significant strengths not available to calculators. These include much larger memories, colour displays, appropriate software and telecommunications capabilities. These attributes are very helpful for applications of mathematics, and it seems highly unlikely these days that professionals engaged in applying mathematics to real world problems would do so without significant ICT use. As computers are rapidly becoming familiar household devices, not just for industrial or professional use, attention to using them well in the mathematics curriculum would seem justified.

General purpose software such as Microsoft's *Word*, *PowerPoint* and *Excel*, has the advantage of being widely available (as it is often bundled with computer purchases) and also widely used outside schools (in businesses, for example). Open Source versions of software of these kinds are also widely available now. Although not designed for mostly mathematical purposes, and certainly not originally intended for educational uses, software of these kinds can be used by primary school pupils to handle tasks involving applications of mathematics. A spreadsheet can be used for data handling and graphing, such as might be helpful for a class project using mathematics to help with planning a school fair. Word processors and presentation software will help pupils write about and present their work to fellow pupils or to others. If applications of mathematics are to move beyond 'exercises' for pupils and become a more prominent part of pupils' work (for example, in the form of projects or activities), computer software of this kind will allow the pupils to consider the key elements of communicating and justifying their findings to interested audiences.

In recent years, apart from software to develop and practice skills (which is not considered here), some mathematical software has been developed explicitly for educational use, most of it for secondary pupils. An outstanding example relevant to the primary school is *Tinkerplots*, a statistics package developed by Key Curriculum Press (2009). Because the software has been designed for educational use, associated materials of various kinds are available for teachers, which is an important practical consideration. An example of *Tinkerplots* use is given later in this chapter.

2.5 The Internet

The Internet has now become a prominent part of people's lives in many countries, certainly including both Singapore and Australia. Consequently, there are many ways in which the Internet might be used for educational purposes, with a typology of these described by Kissane (2009a); the author maintains structured links to suitable websites for mathematics pupils and their teachers (Kissane, 2009b).

As far as application of mathematics in the primary school is concerned, the Internet seems to offer three different kinds of support for pupils. Firstly, some websites offer access to mathematical tools, especially tools for data analysis and simulation. Secondly, real world data can be accessed and in some cases even downloaded, to allow pupils to engage in empirical work; web quests are a particular example of this, but certainly not the only example. Thirdly, the Internet can help pupils to learn about applications of mathematics by reading examples. Examples of all of these opportunities are given later in the chapter.

3 Applying Mathematics with Calculators

As their name suggests, calculators are helpful for doing calculations. Access to calculators by primary pupils will allow much more emphasis than previously on real-world applications of mathematics, for which numbers are not restricted to those that pupils can handle mentally or by hand. When numerical data go beyond what can be comfortably handled mentally, when many calculations are needed, or when more precision is appropriate than good approximations can provide, use of a calculator for computational help seems both unavoidable and desirable. Previous restrictions of applications of mathematics to sanitised data may have exaggerated a common pupil view that mathematics is not really helpful for understanding the world, so that a change towards some calculator use is likely to be significant.

As an example, consider the relationships between the lengths of the sides of a rectangle and its perimeter and area. Once pupils are familiar with these relationships, they are usually provided with exercises involving their use, based on virtual objects drawn or described in textbooks, with measurements provided. Although this might be regarded as an 'application' of the mathematics, it is barely described by the modelling depicted in Figure 1, as it involves little more than recognition and routine use of a relationship. A calculator might be useful when the measurements described are not readily manipulated. When pupil attention is turned from measurement of virtual objects in a textbook to real objects, complete with all necessary information, the situation changes however. Consider the difference between finding the areas of the two rectangles shown in Figure 3, for example. The second task requires pupils to decide what to measure, choose a level of accuracy, determine an area based on the relationship and then consider the reasonableness of the result (and probably then decide to measure again more accurately). It also requires a recognition that measurements are *never* accurate (although textbook authors rarely seem to appreciate that), and that applying mathematics needs to take such things into account.

The objects in Figure 3 are not drawn to scale, but suppose that a pupil measured the sides of the rectangle on the right as 1.5 cm and

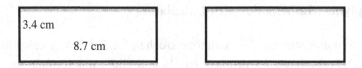

Figure 3. Two different applications of mathematics to find an area

3.7 cm respectively. A first approximation to the area might be regarded as 1.5 x 3.7 = 5.55 sq cm. If measurements are recognised as accurate only to the nearest 0.1 cm however, a better understanding of the area comes from realising that the width could be as small as 1.45 cm or as large as 1.55 cm. This kind of thinking suggests that the area is in fact 1.5 ± 0.05 cm x 3.7 ± 0.05 cm, suggesting that it is likely to be somewhere in the range from 5.29 sq cm to 5.81 sq cm, which is a substantial range. Without access to calculators, such important thinking is rarely undertaken.

Consider further the task of finding the volume of an actual 'real-world' object, such as a building brick, a refrigerator or a classroom. Applying mathematics in such a context is an important activity, from which pupils can learn much about both measurement and about applications. To do so without access to a calculator to handle the unavoidably messy numbers involved would seem hard to defend.

Space precludes continuing with detailed examples, but tasks other than measurement will require primary pupils to undertake arithmetic for which a calculator would seem a sensible tool. Finding the average of a set of data, dealing with large numbers (such as determining how long a billion seconds is or estimating how many times their heart is likely to beat over their lifetime) or finding square roots and cube roots in order to design objects to certain specifications are all examples of this kind.

Pupils need help and experience to use calculators as efficient tools for computation. For example, the *Statistics Singapore* website (Singapore Government, 2009) reports that the population of Singapore is 4987.6 ('000) in mid-2009 with an annual growth rate of 3.1%. There are a number of ways in which pupils might use a calculator with this realistic information to predict Singapore's population in mid-2010. Three of these possibilities are shown in Figure 4.

Figure 4. Three approaches to estimating Singapore's 2010 population

While the first of these might seem 'natural' to pupils, it is incorrect (on this calculator, although not necessarily on others) as the calculator has interpreted 3.1% (correctly) as 0.031 and not calculated 3.1% of the population; this example illustrates that pupils need help to understand exactly how their calculator handles instructions. The second example is incomplete, requiring pupils to next add the increase of 154 616 to the existing population; pupils will need help to use their calculator more efficiently than this. The third screen shows a sophisticated translation of a 3.1% increase to a multiplier of 1.031 and an apparently relaxed approach to leaving numbers in millions, thus requiring an interpretation of the result as an estimated population of 5 142 216 in the middle of 2010; pupils will need the support of teachers to use a calculator efficiently like this, and will incidentally strengthen their understanding of the mathematical ideas involved.

4 Using ICT for Statistics and Probability

Applications of mathematics frequently rely on data, both for building and for verifying models, which is why statistics and probability have become prominent parts of school mathematics curricula in recent years. The use of ICT has revolutionised the practice of statistics, so that data analysis without ICT is very hard to defend. While calculators can handle some aspects of data analysis quite well, dedicated computer software is usually more powerful and more flexible, although is generally not developed for school pupils. As well as software, recent years have seen the development and availability via the Internet of various learning objects, which are helpful tools for applying both statistics and probability.

4.1 Computer software

In recent years, educational software for data analysis has been developed, with the best example relevant to the primary years being *Tinkerplots* (Key Curriculum Press, 2009). *Tinkerplots* is a software package that provides pupils with powerful ways of engaging in data analysis, with the aim of understanding the data well enough to 'tell the story' associated with them. This is especially important for analysing data obtained by pupils themselves to answer questions of their own interest, which is an important reason for attention to applications of mathematics in the curriculum.

An excellent example of this concerns data regarding cats, which are popular household pets in many countries, including Singapore. Some data regarding a set of cats is provided as a demonstration data set for use with the software, and is reproduced here for illustrative purposes. (Readers can download evaluation versions of the software, including the associated data files, from the web link provided by Key Curriculum Press (2009).)

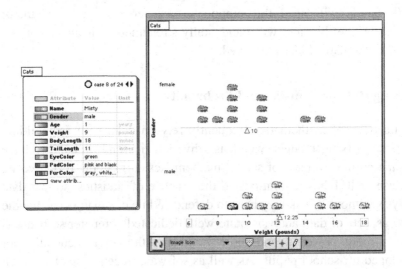

Figure 5. Exploring *Cats* (Key Curriculum Press, 2009)

Figure 5 shows that the data comprise a series of cases, with each cat being a single case, and each case providing information on various attributes of the cat such as gender, age, weight, length, and so on. This resource allows pupils to both ask and find out about several different questions regarding the cats in the set, and even to think about cats more widely. The role of the software is to facilitate data analysis, so that pupils are not distracted by the tasks of calculating statistics or drawing graphs, but instead can focus on the meanings behind the data. Such a role is of course critical for young pupils. Sorting the cases is a primal act of analysis, and the software provides very intuitive ways for pupils to do this to reveal important relationships. It is hard to describe this in words, so that readers are encouraged to try it for themselves via the links provided. The use of *Tinkerplots* allows for less emphasis on sophisticated quantitative analyses and more emphasis on the *story* of the data. The story can be constructed by choosing suitable ways of representing the data visually and numerically. Figure 5 shows one representation of the data to consider the connections between the gender and the body weights of the cats; other representations of the data of course are possible, and relatively easy for pupils to make for themselves, using the software.

It looks as if male cats are generally heavier than female cats, confirmed by the mean body weights (shown in the graph as 12.25 and 10 pounds respectively). While pupils might be able to reach that conclusion, given the set of body weights and a calculator, the use of the software not only facilitates such (time-consuming) computation, but also allows pupils to explore the situation further. In this case, the graph allows and helps pupils to see that there are female cats heavier than most males and that there are also male cats lighter than most females, suggesting that a simplistic conclusion that 'male cats are heavier on average' misses some of the (inevitable) complexity of real world relationships. Further exploration allows pupils to see the extent to which their observations are explained by and related to the ages of the cats and their physical size (e.g. body lengths). Software of this kind provides an opportunity for young pupils to begin to think like a statistician, leaving the complexity of calculations and graphing to the software, and focussing attention on suitable ways to think about the data and interpretations of the graphs and statistics they generate.

Data analysis is most powerful when pupils are addressing problems that are of personal interest to them. They can design and undertake their own data collection, then use software such as *Tinkerplots* to store and analyse their own data. Use of the software does not require high-level ICT skills, and pupils will incidentally learn important lessons associated with statistics such as data editing, dealing with missing cases, identifying and correcting entry errors, and so on.

4.2 *Internet applets for data analysis*

The use of ICT for supporting data analysis allows for attention to be paid to the meaning, rather than to the mechanics of producing representations such as graphs. Some websites offer good support for pupils to produce graphical representations that might otherwise be too hard for pupils to do by hand. Figure 6 shows an example from the *National Library of Virtual Manipulatives* (Utah State University, 2009).

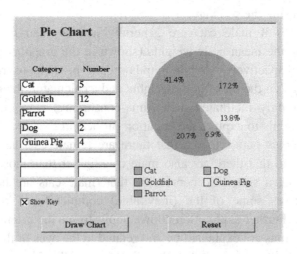

Figure 6. Using the *National Library of Virtual Manipulatives* to make a pie chart

Pie charts are a useful form of data analysis, but quite problematic for pupils (especially young pupils) to draw. A Java applet of this kind allows them to make progress with their data analysis without being lost in the mechanics of finding ratios and measuring angles. Similar applets are available on the same website for drawing bar charts and histograms; there are applets of these kinds also on other sites, such as the excellent NCTM *Illuminations* site. Software of these kinds also provides motivation for pupils to develop computer skills (to take a snapshot of a screen) so that they can add their graphs to presentations in Microsoft *Word* or *PowerPoint*.

4.3 *Simulation via the Internet*

While many practical applications of mathematics rely on accessing authentic data, there are others for which the simulation of random data is more appropriate; these are well handled by the use of ICT tools involving simulation available on the Internet. In Kissane's (2009b) typology, some of these take the form of 'Interactive opportunities', for which links are provided. For example, as well as applets for data analysis, the *National Library of Virtual Manipulatives* provides several applets for pupils to use for simulation, allowing data to be generated, organised and analysed efficiently. These might help pupils to see how randomness can be understood and used for practical purposes. Similarly, a search of the NCTM *Illuminations* site in the Data Analysis and Probability area provides a rich set of resources.

Even before pupils formally study probability, simulation activities can help provide a sound intuitive sense to random activities. If sufficient data are simulated, pupils can begin to see (and indeed to expect) predictable outcomes. Figure 7 shows a tool of this kind from the *Illuminations* website (NCTM, 2009), randomly drawing numbered cards from the set of 21, shown in the top of the diagram.

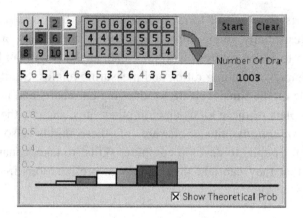

Figure 7. Illuminations applet for Random Drawing (NCTM, 2009)

When a tool of this kind is left to run for some time, remarkable and predictable results are obtained, although in the short term of a few random draws, unexpected events can occur. Experience like this provides a solid base for pupils to begin to apply mathematical thinking to chance events, recognising that mathematics can make good sense of long-term behaviour.

Similarly, the remarkable *Nrich* site in the UK has many simulation tools among its regular resources for pupils and teachers. An important kind of resource offered by *Nrich* are teacher and pupil 'packages', which are collections of related resources that can be downloaded for offline use, particularly helpful when access to the Internet is not readily available in a classroom. For example, one of these (Nrich, 2009a) concerns probability, and offers a set of Flash applets that can be used by the whole class (e.g., on an interactive whiteboard or a whole class display) or used by small groups or individual pupils. Figure 8 shows one of these applets, allowing for efficiently tossing and recording a pair of discs, differently coloured on each side.

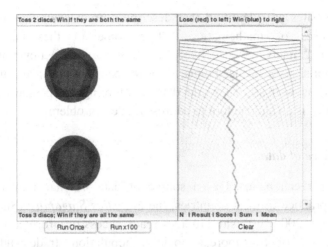

Figure 8. Using the *Nrich* Probability Package to simulate and record data

Successive results are recorded powerfully (using a sheet originally developed by the DIME project). Pupils will see again that, while short term results are unpredictable, in the long run, about half the tosses are the same and about half different, a useful precursor to later, more formal, study of probability.

5 Using the Internet

The Internet has become one of the most important aspects of ICT in recent years, and is now a significant part of the everyday life of many Singaporeans, including primary school pupils. Kissane (2009a) explores some of the ways in which the Internet is significant for both pupils and teachers of mathematics, several of which are of particular importance for applications of mathematics, some examples of which are in the preceding section. Kissane (2009b) systematically catalogues some of the ways in which pupils might learn about mathematics through use of the Internet. For example, one category of use involves 'Interactive opportunity', some examples of which were provided in the previous section. Another category involves 'Reading interesting materials', in which websites are identified that describe current examples of many modern uses of mathematics. While some of these are quite sophisticated, many are also likely to be of interest to upper primary school pupils. As a further bonus, websites can provide recent material for pupils, as well as real world data, reinforcing the perspective that mathematics is a valuable tool to address modern problems.

5.1 *A source of data*

The Internet can be used as a source of data. As has now become common practice in most countries, the *Statistics Singapore* (Singapore Government, 2009) web site provides ready access to information about many aspects of Singaporean society, population, trade and other matters, and was used to obtain population data used earlier in Figure 4. Most data on the Internet are not case-based data, but are summary data, which often limits opportunity for pupil exploration.

Some data are available on the Internet for direct downloading into software packages like *Tinkerplots*. The *Tinkerplots* website itself provides good links to various data sources, although these tend to be a little culturally biased towards the US context, not surprisingly. With such materials, contexts of pupil interest need to be considered, as many of the individual data sets on these websites are not intended directly for school pupils. There would seem to be a good case for teachers to share data sources in this way, to supplement pupils collecting their own data for modelling purposes.

Data are also a key aspect of the concept of 'web quests', which are inquiry-oriented activities in which some or all of the information that pupils interact with comes from resources on the Internet. In the case of mathematics, these usually take the form of a guided exploration of a topic, requiring pupils (or usually groups of pupils) to undertake some web-based research and to generate a product of some kind (such as a report). A good discussion of possible uses of web quests is provided by McCoy (2005). Links to some examples of web quests are provided by Kissane (2009b), but care is needed when using web quests designed by others to make sure that the pupils are directed to appropriate sites (such as those that reflect data related to Singapore).

5.2 *Examples of applications of mathematics*

Some websites describe ways in which mathematics has been applied to the world to solve problems or merely to understand things better. Information of this kind is common for science in television and newspapers, but is much less so for mathematics. From the perspective of pupils, the Internet is relevant here as a source of examples of applications. There are many good articles written for a range of pupils across the primary school, on the *Nrich* site (Nrich, 2009b), very many of which refer to applications of mathematics. These are characteristically short, suitably illustrated and written in a language style appropriate for younger readers. The website is regularly refreshed with new material and repays regular visits.

It is unwise to restrict attention to materials produced directly for primary level pupils, however; there are many other materials that can serve a useful role. The *Experiencing Mathematics* (2009) exhibition is a good example of what can be available to pupils (and their teachers and their parents) via the Internet. Although this exhibition, with more than 50 interactive experiences, has toured many countries in association with UNESCO, the best prospects for most pupils to have access to it is via the Internet. While the application of mathematics is not the sole focus of the museum, it seems reasonable to expect that pupils will have a richer sense of the very many ways in which mathematics is related to the world, and is used to make sense of the world after engaging with the web site. Like other museums, this one appeals to a wide range of pupils, including those in primary school.

As another example, Ron Knott's Fibonacci website (2009) has a wealth of examples of the ways in which mathematical ideas and mathematical thinking have helped illuminate the world, often in surprising ways. From the simple beginnings of a number sequence in which each term after the second term is the sum of the preceding two terms (1, 1, 2, 3, 5, ...), many wondrous applications of mathematics emerge in both the natural world of plants and animals as well as the human world of arts and architecture. While some of the mathematics involved is beyond the primary school, many of the ideas are accessible to primary pupils, so that a website of this kind will help make clear that mathematics can be applied to many phenomena.

As a third example, the American Mathematical Society (2009) has for some time published various posters showing mathematics applied to many activities. Categories of science, nature, technology and human culture allow pupils to see that mathematics is used to understand and to inform many activities in the world, including those that do not at first seem connected to mathematics. The series also projects the important view that mathematics is relevant to today's world, not only yesterday's world. Because of the format of posters (available in both more and less sophisticated versions) the website also offers an opportunity to obtain materials for classroom walls that help to reflect a range of modern applications of mathematics.

These, and many other examples, are referred to by Kissane (2009b) in the category of 'Reading interesting materials'. The Internet has the potential to allow pupils, including primary school pupils, to explore many of the ways in which mathematics has been applied by others to phenomena in the world. While necessarily there are many of these that are concerned with sophisticated mathematical ideas (including those beyond the immediate purview of the primary school), together they offer an opportunity for mathematics to be recognised by pupils as a significant, powerful and useful force in the modern world, worthy of further study. Perhaps ironically, this view of mathematics is not always evident in school curricula, and is conspicuously absent from media such as television, the movies and popular print media. So, the Internet might serve the useful role of helping pupils to develop a positive attitude towards applications of mathematics and an incentive to learn more about it later. Developing such a perspective is a key part of primary school mathematics education.

6 Conclusion

A range of ICT is now available for pupils, and it is clearly appropriate for the mathematics curriculum to take advantage of this, as the Ministry of Education (2006) has noted. The use of ICT for applications of mathematics is constrained by facilities, the curriculum and the time available. Some examples of ICT (such as calculators) are likely to be available everywhere, and to meet many needs. Other manifestations of ICT, such as specialist software, are more powerful, but usually require more resources to use effectively.

This chapter has suggested that ICT can play a substantial role in improving the statistics and probability components of the school curriculum in particular. In ways that were not previously possible, pupils can engage in genuine data analysis, and explore questions of their own interest. They can also explore chance phenomena through simulation, not previously accessible without access to ICT. The Internet offers both tools for applying mathematics and examples of mathematics being applied, relevant to both primary pupils and their teachers. Extensive backgrounds in the use of ICT are not needed to take advantage of these opportunities.

References

American Mathematical Society. (2009). *Mathematical Moments*. Retrieved November 8, 2009, from http://www.ams.org/mathmoments/

Burkhardt, H. (1981). *The real world and mathematics*. UK, London: Blackie.

Experiencing mathematics (2009). Retrieved on November 26, 2009, from http://www.experiencingmaths.org/

Groves, S. (1997). The effect of long-term calculator use on children's understanding of number: Results from the Calculators in Primary Mathematics project. In N. Scott &

H. Hollingsworth (Eds.), *Mathematics: Creating the future* (pp. 150-158). Australia, Melbourne: The Australian Association of Mathematics Teachers.

Kaput, J. (1992). Technology and mathematics education. In D. A. Grouws (Ed), *Handbook of Research on Mathematics Teaching and Learning* (pp. 515-556). Reston, VA: National Council of Teachers of Mathematics.

Key Curriculum Press. (2009). *Tinkerplots® Dynamic data Exploration.* Retrieved on November 8, 2009, from http://www.keypress.com/x5715.xml

Kissane, B. (1997). Growing up with a calculator. *Australian Primary Mathematics Classroom*, 2(4), 10-14. [Available online at http://wwwstaff.murdoch.edu.au/~kissane/chapters/apmc.pdf]

Kissane, B. (2009a). What does the Internet offer for pupils? In C. Hurst, M. Kemp, B. Kissane, L. Sparrow & T. Spencer (Eds.), *Mathematics: It's Mine: Proceedings of the 22nd Biennial Conference of the Australian Association of Mathematics Teacher* (pp. 135–144). Australia, Adelaide: Australian Association of Mathematics Teachers.

Kissane, B. (2009b). *Learning mathematics on the Internet.* Retrieved on November 25, 2009, from http://wwwstaff.murdoch.edu.au/~kissane/internetmaths.htm.

McCoy, L. P. (2005). Internet WebQuest: A context for mathematics process skills. In J. M. William & P. C. Elliott (Eds.), *Technology-Supported Mathematics Learning Environments: 67th Yearbook* (pp 189-201). Reston, VA: National Council of Teachers of Mathematics.

Ministry of Education. (2006). *Mathematics syllabus – Primary.* Singapore: Author.

NCTM (National Council of Teachers of Mathematics). (2009). *Illuminations: Random Drawing Tool.* Retrieved on November 12, 2009, from http://illuminations.nctm.org/ActivityDetail.aspx?ID=67

Nrich. (2009a). *Experimenting with probability.* Retrieved on November 8, 2009, from http://nrich.maths.org/public/viewer.php?obj_id=5541

Nrich. (2009b). *Articles.* Retrieved on November 13, 2009 from http://nrich.maths.org/5547

Knott, R. (2009). *Fibonacci numbers, the golden section and the golden string.* Retrieved November 26, 2009, from http://maths.surrey.ac.uk/hosted-sites/R/Knott/Finonacci/

Singapore Government. (2009). *Statistics Singapore.* Retrieved on November 26, 2009, from http://www.singstat.gov.sg/

Utah State University. (2009). *National Library of Virtual Manipulatives.* Retrieved on November 26, 2009, from http://nlvm.usu.edu/

Chapter 4

Application Problems in Primary School Mathematics

YEO Kai Kow Joseph

This chapter describes the characteristics of application problems and the thinking processes involved in solving application problems at the primary level. Four examples of application problems are used to show the benefits of infusing such problems into primary mathematics lessons.

1 Introduction

In line with global trends in mathematics education, there has been an increasing emphasis on the application of mathematics including the solving of problems posed in realistic situations (Cockcroft, 1982; NCTM, 1991; Gravemeijer, 1994; Dowling, 1991; Lowrie & Logan, 2006; Gainsburg, 2008). A range of propositions has been made. Some suggest that primary school pupils develop their understanding of mathematics by applying it to textually represented realistic problems; others advocate for the application of mathematics to real problems outside the mathematics classroom. However, in many countries, high stakes assessment in mathematics has continued to use short stereotypical problems which set mathematical operations within limited and artificial contexts (Cooper & Harries, 2002). These short stereotypical types of problems expect pupils to be able to apply and practise a recently acquired algorithm (Kulm, 1994). There is a concern that this expectation encourages teachers to "load" pupils with a knowledge base of rules, algorithms and formulae as they would with machines and expect the pupils to have learnt, to be able to regurgitate the formulae and solve problems faultlessly and with consistent solutions each time. Not only has this resulted in pupils listening and absorbing knowledge passively but it has also lead pupils into developing mathematics avoidance (Collin, Brown, & Newman, 1989). Although the latest primary school mathematics textbook series in Singapore has introduced many new types of mathematics problems as examples and exercise questions, some pupils may not appreciate how mathematics could be applied in daily life. Therefore, mathematical problems put into context would assist the pupils to value the real life significance of mathematics concepts.

The primary aim of teaching mathematics in Singapore is to enable pupils to solve problems (Ministry of Education [MOE], 2006). The attainment of this aim is dependent on five factors; specifically, concepts, skills, attitudes, metacognition and processes. Processes include thinking skills and heuristics involved in solving problems. Thinking skills such as classifying, comparing, sequencing, analysing parts and wholes, identifying patterns and relationships, induction, deduction, and spatial visualisation (MOE) are taught to pupils. One of the major revisions made to the Singapore mathematics curriculum framework with effect from 2007 is the change in the processes components, such as "reasoning, communication and connections" and "applications and modelling" being part of processes. These changes reinforce implementation of Thinking Programme (MOE, 1998) in Singapore schools that indicated a greater emphasis on infusing thinking skills in the primary mathematics lessons.

Despite widespread recognition of the importance of connecting school mathematics to the real world, little is known about how and how often it actually happens in classrooms. The few relevant studies (e.g., Verschaffel et al., 1999; English & Watters, 2005) suggest that the use of application problems in mathematics teaching is well below the level and quality envisioned by scholars. Moreover, most primary school pupils believe that mathematics is a collection of facts and skills to be mastered. However, mathematics is also about thinking and making sense of the authentic situations through numbers, measurement and symbols. Thus, in this chapter, the main purpose is to review the characteristics of application problems and the thinking process involved in solving application problems. This chapter also describes four application problems that can be infused in the teaching and learning of mathematics at the primary level.

2 Characteristics of Application Problems

Within the Singapore mathematical curriculum framework problems are both a means and an end in school mathematics instruction. Problems have been classified differently by various researchers. A simple classification by Charles and Lester (1982) identified six types of problems; namely:
  1. drill exercises,
  2. simple translation problems,
  3. complex translation problems,
  4. process problems,
  5. puzzles, and
  6. applied problems.

Applied problems required the pupils to use a wide variety of mathematics skills, processes and facts to solve realistic and everyday problems. Subsequently, Charles, Lester and O'Daffer (1987) further refined and categorised the problems into four types:
  1. one-step problems,
  2. multiple-step problems,
  3. process problems, and
  4. applied problems.

They further refined applied problems as 'situational problems' that require the problem solver to collect data outside the problem statement. Blum and Niss (1991) further elaborated that the applied mathematical problem should belong to some segment of the real world situations where mathematical concepts, methods, and results were involved. There are some researchers who would use the term applied problems and application problems interchangeably. For instance, an application problem should be a situation where mathematical knowledge is launched, activated, applied and evaluated within the problem embedded in real world settings (Stillman & Galbraith, 1998). Moreover, "using mathematics to solve real world problems...is often called *applying* mathematics, and a real world problem which has been addressed by means of mathematics is called an *application* of mathematics" (Niss, Blum, & Galbraith, 2007, p. 10). In fact, Galbraith (1999) further explained that although the mathematics and context are related in an application of mathematics, they are separable. It meant that after applying the necessary mathematics to solve the problem in some given context, the context may not be "required" any more. Whatever, the views and differences in definition, we need to be mindful that application problems has to have some link with real life situation.

Providing pupils with opportunities to solve 'good' application problems should be the main consideration in mathematics instructions. Clarke and McDonough (1989) described a good problem as one that "requires the learner to review all relevant knowledge, to make decisions, to select, to combine, to plan to evaluate, and to communicate" (p. 3). According to Schoenfeld (1994), a good problem should be one that can be extended to lead to mathematical explorations and generalisations to encourage mathematical thinking. Such problems can encourage pupils to represent the information in various ways, for example, symbolically, numerically, pictorially or graphically. Typically, these problems also have more than one way of solving them. Usually, the solutions may require interpretation and judgment before understanding can take place. For understanding to surface, Carpenter and Lehrer (1999) mentioned that mathematical problem-solving processes in the classroom should provide pupils with tasks that can activate the following five forms of interrelated mental activities:

1. develop appropriate relationships,
2. extend and apply their mathematical knowledge,
3. reflect about their own experiences,
4. articulate what they know, and
5. make mathematical knowledge their own.

In other words, utilising application problems in the mathematics lesson, enables opportunities for the mathematics teacher and pupils not only focus on the reasonableness of the final answer. Rather application problem activity encourages the focus to be more directed to the mathematical concepts and skills and the relevance of mathematics and its importance in real life. Where appropriate, application problems also serve as contexts for pupils to acquire thinking skills.

3 Thinking Process of Solving Application Problems

In contrast to traditional mathematics activities, Lester (1989) noted that solving problems that arise or are formulated in everyday situations often require "mathematical procedures and thinking processes that are quite different from those learned in school. Furthermore, people's everyday mathematics often reflects a higher level of thinking than is typically expected or accomplished in school" (p. 33). Such observations have led researchers such as Lesh and Harel (2003) to urge that the problem solving situations that should be emphasised in the classroom should include simulations of real-life experiences. Howe and Warren (1989) claimed that higher-order thinking skills cannot be learnt well unless teachers emphasize it and use it on a continuing basis. Howe and Warren also pointed out that the availability of ample amounts of pupils' active participation is an important factor to enhance the development of pupils' higher-order thinking skills. To achieve higher-order thinking, pupils should be involved in the transformation of knowledge and understanding. This includes creating a communicating environment for pupils' effective interaction, encouraging them to verify, question, criticize and assess others' arguments, engaging in constructing knowledge through various processes and generating new knowledge through self-exploration. In the pupils' perspectives, they also need to be aware that they must be an active learner taking initiatives and responsibilities in their own learning. In fact, for many years, studies have shown that higher-order thinking can be taught, unlocked, nurtured, and developed (Lumsdaine & Lumsdaine, 1994). Similarly, in a recent study, Stein, Engle, Smith and Hughes (2008) emphasised that implementing a cognitively demanding instructional mathematical task with multiple possible solutions should be supported by teachers' understanding of their pupils' current mathematical thinking and practices. Therefore, as long as teachers consciously teach thinking skills and provide opportunities for interaction, the implementation and practice of higher-order thinking can be greatly improved in our mathematics classrooms. Here, thinking skills such as "classifying, comparing, sequencing, analysing parts and whole, identifying patterns and relationships, induction, deduction and spatial visualisation" (MOE, 2006) could be introduced into the mathematics lessons.

When pupils are engaged in solving mathematical problems, 'thinking' will take them through the learning process. However, our primary school pupils may not be consciously aware of these thinking skills that they have always been using. Thinking skills and heuristics form one of the five interrelated components on which the attainment of mathematical problem solving ability is dependent on (MOE, 2006). The syllabus states that "thinking is inherent in the subject." (MOE, 2000, p. 17). It indicates that the development of thinking skills and mathematics learning should not be taught separately. Instead, thinking skills should be consciously infused and reinforced in daily mathematics classroom teaching. Because "thinking skills should be consciously infused and reinforced through the learning of mathematical concepts and problem solving" (MOE, 2000, p. 17), it can be inferred that activities fostering thinking are included in the curriculum albeit they are not listed overtly as such. In the Singapore Revised Mathematics Syllabus (MOE, 2006), one of the main emphases of the primary level mathematics curriculum has been the acquisition and application of mathematical concepts and skills for everyday life. While the revised curriculum continues to emphasise this, there is now an even greater focus on the development of pupils' abilities to conjecture, discover, reason and communicate mathematics with the use of application problem. The appropriate use of application problem in the classroom is the key factor.

As aforementioned, the use of application problems in meaningful contexts is often claimed to be enriching for pupils as their mathematical experiences become connected to real life experiences. Moreover, application problems also have the added benefit of helping pupils grasp concepts through linking abstract, unfamiliar mathematical concepts to real life situations (Yeap & Kaur, 1992). Although application problems have their value in the learning mathematics for the primary levels, we should not advocate using them merely because they are popular. Instead, teachers need to establish thoughtful rationales for deciding how and when to use application problems in their classrooms. Undeniably, mathematics educators and teachers would agree that lessons involving application problems tend to emphasise more on pupils' conceptual understanding rather than rote learning. Therefore schools should strongly encourage the use of application problems in all aspects of mathematical instruction including the development of mathematical concepts and the acquisition of computational skills.

4 Sample Application Problems for Primary School Pupils

Since an application problem is an effective instructional tool for enhancement of mathematical concepts and improving thinking process, it will be useful to provide four such appropriate application problems for primary mathematics teachers to infuse in their mathematics lessons. In my work with primary mathematics school teachers in Singapore, I frequently hear the concerns about infusing thinking skills in mathematical problems. Amongst others, primary mathematics school teachers also wish that their school pupils to develop the following thinking skills while solving application problems: analysing parts and whole, comparing, identifying patterns and relationship as well as spatial visualisation. The following section describes four application problems that could be infused in the teaching and learning of mathematics at the primary level.

4.1 *Analysing parts and whole*

Analysing parts and whole in an application problem is to recognise and articulate the parts that together constitute a whole (MOE, 2000). In a traditional classroom, primary school pupils may have little opportunity to explain and justify the mathematical process involved in their mathematical solutions. Sometimes they may not be able to understand what is expected in explaining their thinking. Although they may

> **Application Problem 1: Percentage (Primary 6)**
> Mr Lim decides to buy a new car which is priced in the showroom for $64,500.
> (a) If he pays in cash, he will be given a discount.
>   Calculate the percentage discount given if he only needs to pay $59,340 in cash.
> (b) If he offers his old car in part exchange, the salesman is willing to allow him $30,900 towards the cost of the new car. Mr Lim borrows the remaining amount from a finance company which charges simple interest at the rate of 6.5% per annum. If he takes a 4-year loan from the finance company, calculate the
>   i) total interest he has to pay.
>   ii) amount he has to pay each month when he spreads his repayment of the loan equally for each month over the 4 years.
> (c) If the salesman is able to sell the car at $64,500, he will have made a profit of 20% on the cost price. Calculate the cost price of each new car.
> (d) The salesman had made 5 sales earlier that day at an average price of $58,350 each. If Mr Lim pays $64,500 for his car, find the average price of each of the cars the salesman would have sold that day.

perform certain computations, they do not know how to explain why they do them or why they work. Even if the teacher insists that the pupils explain and justify, they may simply mimic what the teacher has said in class. Problems like Application problem 1, can be used to provide pupils opportunities to identify and explain how each part of the computation contributes to the whole. In this case, the whole refers to the $64 500 which is the price of the new car. This entire process of reasoning involves the flexibility in thinking about the transactions and amount that is related to the whole ($64 500). Within the problem, the requirement that pupils to make informed judgments and decisions regarding the use and management of money provides an opportunity for pupils to explore and appreciate financial literacy. Increasing pupils' awareness of how mathematical reasoning can assist authentic day-to-day situational decision-making of money matters is an important learning outcome. Many people in their daily life need to deal with various problems involving money such as buying lunch, depositing money in a bank and getting a loan to buy a car.

4.2 Comparing

To compare is to use common attributes to identify commonalities and discrepancies across numerous sets of information (MOE, 2000). For instance, in problem 2, pupils are required to calculate and compare the cost of covering the kitchen floor with Type X tiles with the cost of covering the kitchen floor using the combination of Type Y and Z tiles. Moreover, through solving the problem, the concept of area of rectangle is reinforced. For pupils who are lacking in conceptual knowledge of area of composite figure, they would have a good grasp of 'breaking up' a composite figure especially when they are finding the area using the Type Y and Z tiles. Pupils also need to study the similarities and difference between Type X, Y and Z tiles so as to fit it into a rectangular kitchen floor. Compare this application problem to the standard textbook problem of asking the pupils to find the area of composite figures, application problem 2 helps the pupils to relate concept of area of composite figures in a meaningful context. Although pupils are able to find the cost and the number of Type X, Y and Z tiles covering the kitchen floor, they have to make realistic assumptions and decisions that would require their reasoning in order to find the answers. Furthermore, when faced with such an application problem, pupils should somehow represent its structure by identifying the quantities and the relationships between them in order to make a decision and justification.

Application Problem 2: Area of composite figures (Primary 4)
A house owner wishes to tile his rectangular kitchen floor which is 4 m long and 3 m wide. Rectangular floor tiles are sold in the following sizes:

Type of tile	Size	Cost per tile
X	50 cm by 50 cm	$32.50
Y	30 cm by 30 cm	$ 8.20
Z	10 cm by 20 cm	$ 3.70

(a) Calculate the cost of covering the floor with Type X tiles.
(b) (i) Calculate the largest kitchen floor area which can be covered using whole tiles of Type Y.
    (ii) Hence calculate the number of whole tiles of Type Z required to complete the covering of the floor.
(c) Calculate the amount of money that the house owner would have saved if he decides to cover his kitchen floor using the above combination of Type Y and Z tiles as compared to covering the floor with Type X tiles only.

72 *Mathematical Applications and Modelling*

4.3 *Identifying patterns and relationship*

To identify patterns and relationship is to recognise the specific variations between two or more attributes in a relationship that yields a reliable or repeated scheme (MOE, 2000). In order for the pupils to examine patterns and note regularities in a set of values, application problem 3 involves examining patterns in amount of money collected over 20 days. At the first glance of application problem 3, it appears Mary would receive more if she chooses to collect $4000 first. To identify the patterns and relationship, pupils could calculate the first five cases so as to study the growth of the money. Typically, pupils could display the growing amount by recording a pattern in a T-table where the first column shows the number of days and the second column shows the amount paid. Teacher can encourage pupils to describe and analyse the relationship that exits for the first five total amounts. Their explanations can be tested by asking pupils to predict what the next result would be. The aim is to have pupils see a pattern that they can generalise into a rule. As pupils become more proficient with verbalising a rule, the teacher could show them how the rule can be written using symbolic notation with letters from the T-table and numbers. The teacher could also get the pupils to sum up the amount of money received for each five-day to facilitate the calculation of the grand total. Pupils' reaction to the fact that the total amount collected is $524288 over the 20 days is typically one of the amazement and utter disbrief. This is because primary school pupils do not realise the amount of money could grow exponentially. This has certainly created an opportunity for pupils to explore, appreciate number patterns, and 'see' the relationship in terms of its power of 2.

Application Problem 3: Whole Numbers (Primary 5)
One month ago, Mary bought a charity draw ticket. She was lucky because she won! The only dilemma was that she had to choose between 2 prizes. Four thousand dollars all at once or 1 cent on the first day, 2 cents on the second day, 4 cents on the third day, 8 cents on the fourth day, 16 cents on the fifth day and so on for 20 days in a row.
Which prize should she choose?
How can you show that it is the better choice?

4.4 Spatial visualization

Spatial concepts, ideas of shapes and sizes are part of the experience of all pupils from their early school years. This includes visualising a situation or an object and mentally manipulating various alternatives for solving a problem related to a situation or object without benefit of concrete manipulative (MOE, 2000). In other words, pupils from a young age need to develop spatial sense. Spatial sense involves both visualisation and orientation factors (Owens, 1990). Visualisation is the ability to mentally portray how shapes or objects emerge under some rigid motion or transformation (Robison, 2006). Robison further added that orientation includes the ability to figure out positions of shapes or objects and to maintain an accurate perception of the shapes or objects under different orientations. Introducing these ideas of informal geometry will assist the pupils to realise and appreciate that mathematics is more than numbers and computation.

For primary school teachers wishing their pupils to discover spatial properties and relationship that connect to their real world, tessellation offers a good starting point. Tessellation, a topic taught in primary four geometry strand in the revised mathematics syllabus in Singapore lends itself to learning informal geometry through exploring and discovering properties of shapes and their relationships with respect to sides, angles, symmetry, translation, rotation, reflection and other geometrical properties. In Application problem 4, pupils can be guided through discovery process by exploring the various ways to cover a surface completely using physical representations of the given shapes. The physical manipulation of shapes supports pupils to understand and explain why some figures tessellate and why others do not. Through a whole class discussion report back the teacher can then collate pupils' thoughts about the necessary conditions for tessellation. Teaching tessellation in a mathematics class can reveal beautiful geometric relationships and patterns. Solving application problem 4 involving tessellation allows pupils to develop their own artistic creativity by applying their mathematical knowledge and makes them more aware of the use of geometric patterns in the real world.

Application Problem 4: Tessellation (Primary 4)
Mr Yeo sells window panes in his decor shop in Ang Mo Kio. One day, Mr and Mrs Lee came to Mr Yeo's shop to purchase some window panes. Mr Yeo showed them the following panes and told them that only some of these panes could be used to create window panes without leaving any gaps between the panes.

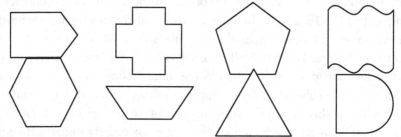

Mr and Mrs Lee would like to choose only one type of panes for their windows in the bed room and two different types of panes for their windows in the living room. Help Mr and Mrs Lee to find the suitable window panes.

These four application problems exemplify how application problems help primary school pupils to explore and apply various types of mathematical concepts. The different application problems afford a range of rich learning experience that foster thinking skills. The four application problems exemplify to make the use of application problems in the classroom a meaningful one where emphasis is on the process (reasoning and thinking) rather than the product (final answer). Pupils could make use of thinking skills that they have learnt and applied them to solve mathematical application problems systematically. Application problems enables mathematics teachers and pupils to be involved in solving mathematical problems go beyond computations and rote algorithms. Mathematics is undoubtedly a very crucial examination subject but mathematical knowledge and thinking should be subsumed under the overall mathematics curriculum and instruction.

5 Conclusion

The thinking skills emphasised above are significant for any pupil learning mathematics. These thinking skills have to be fostered over a long period of time through the use of carefully chosen application problems by the mathematics teachers. A single application problem cannot reveal all the thinking skills; rather we need to draw on a range of application problems spread over various topic areas. The four application problems can provide mathematics teachers access to pupils' thinking and conceptual understanding (Caroll, 1999). Teachers who assign real-world application problems must ensure that they use authentic data. They are no more time-consuming to correct than the homework exercise that teachers usually give. When used regularly, the pupils can develop the skills of reasoning and communication in words and diagrams. Pupils in presenting their solutions to others could compare and examine each other's method. It is an approach that enables primary school mathematics teachers to teach mathematics that aligns with the spirit and intent of "Teach Less, Learn More" (TLLM). The focus of TLLM is on thinking, reasoning and engage learning that characterise the shift from practicing isolated skills towards developing rich network of conceptual understanding. In addition, to help pupils accustom themselves with the use of application problems; assessment needs to be modified at the school level in order to focus on this new development. However, there is a need to strike a balance between basic numeracy skills, conceptual understanding and problem solving. The acquisition of thinking skills would help pupils in the learning of mathematics, and thus allow a higher level of appreciation of mathematics among our pupils. Ultimately, we hope to develop in our pupils positive habits that will assist them to become critical, creative and self-regulated thinking learners.

References

Blum, W., & Niss, M. (1991). Applied mathematical problem solving, modelling, applications, and links to other subjects — State, trends and issues in mathematics instruction. *Educational Studies in Mathematics, 22*(1), 37-68.

Carpenter, T, P., & Lehrer, R. (1999). Teaching and learning mathematics with understanding. In E. Fennema & T. A. Romberg (Eds.), *Mathematics classrooms that promote understanding* (pp. 19-33). Mahwah, NJ: LEA.

Carroll, W. M. (1999). Using short questions to develop and assess reasoning. In L.V. Stiff & R. Curcio (Eds.), *Developing mathematical reasoning in grades K-12, 1999 Yearbook* (pp. 247-255). Reston, Va.: NCTM.

Charles, R. I., & Lester, F. K. (1982). *Problem solving: What, why and how*. Palo Alto, California: Dale Seymour.

Charles, R., Lester, F., & O'Daffer, P. (1987). *How to evaluate progress in problem solving*. Reston, VA: National Council of Teachers of Mathematics.

Clarke, D., & McDonough, A. (1989). The problems of the problem-solving classroom. *The Australian Mathematics Teacher, 45*(3), 20-24.

Cockcroft, W. H. (1982). *Mathematics Counts*. London: HMSO.

Collin, A., Brown, J. S., & Newman, S. E. (1989). Cognitive apprenticeship: Teaching the crafts of readings, writing and mathematics. In L. B. Resnick (Ed.), *Knowing learning and instruction: Essays in honour of Robert Glaser* (pp. 453-494). Hillsdale, NJ: Erlbaum.

Cooper, B., & Harries, T. (2002). Children's responses to contrasting 'realistic' mathematics problems: Just how realistic are children ready to be? *Educational Studies in Mathematics, 49*(1), 1-23.

Dowling, P. (1991). The contextualising of mathematics: Towards a theoretical map. In M. Harris (Ed.), *School, Mathematics and Work.*. Basingstoke: Falmer Press.

Galbraith, P. (1999, July). *Important issues in application and modelling.* Paper presented at the AAMT Virtual Conference 1999, Adelaide, Australia.

Gainsburg, J. (2008). Real-world connections in secondary mathematics teaching. *Journal of Mathematics TeacherEducation,* 11*(3),* 199-219.

Gravemeijer, K. (1994). *Developing Realistic Mathematics Education.* Utrecht: Freudenthal Institute.

Howe, R. W., & Warren, C. R. (1989). *Teaching critical thinking through environment education.* (ERIC Document Reproduction Service No. ED324193).

Kulm, G. (1994). *Mathematics Assessment: What works in the classroom?* San Francisco: Jossey-Bass Publisher.

Lesh, R., & Harel, G. (2003). Problem solving, modelling and local conceptual development. *Mathematical Thinking and Learning, 3*(3), 157-189.

Lester, F. K., Jr. (1989). Mathematical problem solving in and out of school. *Arithmetic Teacher,* 37(3), 33-35.

Lowrie, T., & Logan, T. (2006). Using spatial skills to interpret maps: Problem solving in realistic contexts. *Australian Primary Mathematics Classroom,* 12*(4),* 14-19.

Lumsdaine, E., & Lumsdaine, M. (1994). *Creative problem solving: Thinking skills for a changing world.* New York: McGraw-Hill.

English, L. D., & Watters, J. J. (2005). Mathematical modeling in the early school years. *Mathematics Education Research Journal, 16*(3), 58-79.

Ministry of Education (1998). *Thinking programme: The explicit teaching of thinking. Teachers' manual. Secondary Two.* Singapore: Curriculum Planning and Development Division.

Ministry of Education (2000). *Mathematics syllabus: Primary.* Singapore: Curriculum Planning and Development Division.

Ministry of Education. (2006). *Mathematics syllabus: Primary.* Singapore: Curriculum Planning and Development Division.

NCTM (National Council Of Teachers Of Mathematics). (1991). *Professional Standards for Teaching Mathematics.* Reston, VA: Author.

Niss, M., Blum, W., & Galbraith, P. (2007). Introduction. In W. Blum, P. L. Galbraith, H. W. Henn, & M. Niss (Eds.), *Modelling and applications in mathematics education: The 14th ICMI study* (pp. 3-32). New York: Springer.

Owens, D. T. (1990). Research into practice: Spatial abilities. *Arithmetic Teacher, 37*(6), 48-51.

Robison, S. (2006). Developing geometric thinking and spatial sense. In G. Cathcart., Y.M. Pothier, J.H. Vance, & N. S. Bezuk (Eds.), *Learning mathematics in elementary and middle schools* (4th edition.) (pp. 285-315). Columbus, OH: Merrill Prentice Hall.

Schoenfeld, A. H. (1994). Reflections on doing and teaching mathematics. In A. H. Schoenfeld (Ed.), *Mathematical thinking and problem solving* (pp. 53-70). Hillsdale, NJ: Erlbaum.

Stein, M.K., Engle, R.A., Smith, M.S., & Hughes, E.K. (2008). Orchestrating Productive Mathematical Discussions: Five Practices for Helping Teachers Move Beyond Show and Tell. *Mathematical Thinking and Learning, 10(*4), 313-340.

Stillman, G., & Galbraith, P. L. (1998). Applying mathematics with real world connections: Metacognitive characteristics of secondary students. *Educational Studies in Mathematics, 36*(2), 157-195.

Yeap, B. H., & Kaur, B. (1992). Mathematical problem solving in regular classrooms. *Teaching and Learning, 13*(1), 17-24.

Verschaffel, L., De Corte, E., Lasure, S., Van Vaerenbergh, G., Bogaerts, H., & Ratinckx, E. (1999). Learning to solve mathematical application problems: A design experiment with fifth graders. *Mathematical Thinking and Learning, 1*(3), 195-229.

Chapter 5

Collaborative Problem Solving as Modelling in the Primary Years of Schooling

Judy ANDERSON

Collaborative problem solving which requires the use of processes such as questioning, analysing, reasoning and evaluating to solve particular types of tasks mirrors mathematical modelling. Rich modelling tasks provide an alternative to problem solving and can be used as a catalyst for primary school pupils to develop deeper understanding of mathematical ideas, problem-solving skills and techniques, and collaborative responsibility. This chapter outlines the similarities and differences between modelling and problem solving, presents the characteristics of rich modelling tasks, and describes the importance of collaboration through cooperative groups. Examples of tasks are used to provide advice about the development of modelling tasks as well as the implementation of a modelling approach to problem solving in primary school classrooms.

1 Problem Solving and Modelling – Similarities and Differences

Problem solving is recognised as an important life skill involving a range of processes including analysing, interpreting, reasoning, predicting, evaluating and reflecting. It is either an overarching goal or a fundamental component of the school mathematics curriculum in many countries (e.g., Ministry of Education, 2006; National Curriculum Board, 2009). However, developing successful problem solvers is a complex task requiring a range of skills and dispositions (Stacey, 2005). Pupils need deep mathematical knowledge and general reasoning ability as well as heuristic strategies for solving non-routine problems. It is also necessary to have helpful beliefs and personal attributes for organizing and directing their efforts. Coupled with this, pupils require good communication skills and the ability to work in cooperative groups (Figure 1).

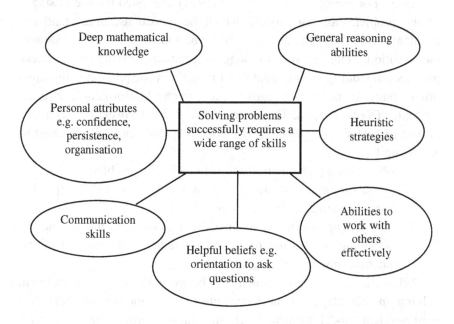

Figure 1. Factors contributing to successful problem solving (Stacey, 2005, p. 342)

While problem solving and modelling require similar skills and dispositions and have characteristics in common, there are important differences, which impact on the types of tasks teachers choose and the way teachers engage pupils in modelling experiences. Zawojewski (2007) asserts problem solving is usually defined with respect to the problem solver and the process of problem solving involves a search for a means to solve the problem, usually with a focus on correct procedures and correct solutions. Whereas in modelling, the nature of the tasks posed is now the focus so that appropriate tasks require interpretation of the information and interpretation of the desired outcomes. This is best achieved in cooperative groups of pupils as they design and identify flaws in proposed models, understand limitations, as well as test and revise the model they choose for the task.

Problem solving and modelling are complex processes but to support pupils' problem solving and modelling efforts, approaches have been developed to guide student thinking and promote metacognitive processes. For problem solving, Polya (1971) suggested the use of stages of the problem solving process which he named as: understand the problem; devise a plan; carry out the plan; and look back and examine the solution. This is not to suggest problem solving is a linear progression from 'givens to goals' but rather a cyclic process with pupils often having to backtrack to earlier stages to check information or refine strategies (Verschaffel & De Corte, 1997, p. 577). Swetz and Hartzler (1991) presented a set of stages for modelling that could also be used to assist pupils:

1. observing a phenomenon and delineating the problem;
2. conjecturing the relationships among factors and interpreting them mathematically (mathematizing);
3. applying appropriate mathematical analysis to the model; and
4. obtaining results and reinterpreting them in the context of the phenomenon (p. 3).

While there are some similarities between the stages of problem solving and the stages of the modelling process, the development of a mathematical model from a situation, which seems to be "seemingly non-mathematical in context", is unique to mathematical modelling (Swetz & Hartzler, 1991, p. 2). The stages of the modelling process have been represented as a cyclic process as shown in Figure 2, although a fifth stage of testing and refining the model could be added.

While there are differences between problem solving and modelling in mathematics, particularly in relation to the types of tasks and the stages of the process, there are clearly similarities with the potential for pupils to develop the important skills and dispositions required for successful problem solving. To summarise, Zawojewski, Lesh and English (2003) describe the link between problem solving and modelling as:

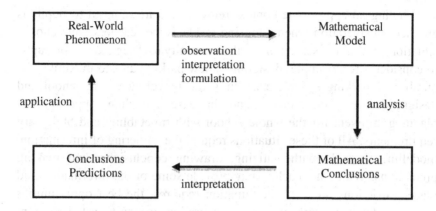

Figure 2. The stages of the modelling process (Swetz and Hartzler, 1991, p. 3)

- teams working on problem situations;
- partitioning a complex situation into parts;
- communicating information; and
- planning, monitoring and assessing immediate results.

These are all worthwhile and desirable outcomes for any mathematics curriculum. Choosing appropriate modelling tasks to support these processes is a key responsibility of teachers.

2 Characteristics of Modelling Tasks

Rich model-eliciting tasks have particular characteristics which involve collaborative problem solving but also require pupils to 'mathematize situations', usually situated within a real-life context. Mathematizing is a critical component of good modelling tasks and may require support for pupils during initial implementation. Tasks must be based on important mathematical ideas so that deeper and higher-order understandings are possible (Lesh et al., 2000). Such tasks have the potential to help teachers recognise and reward pupils with a broader range of mathematical skills and dispositions and allow teachers to gather useful information about pupils' strengths and weaknesses.

Finding an appropriate context relevant and meaningful to pupils is another consideration when choosing tasks. Some teachers have chosen situations within the school or local community to design tasks for pupils to consider – for example: designing a new parking area to maximise the number of parking spaces; examining the recycling in the school and designing ways to improve and increase recycling opportunities; planning an event for the whole school with timetabling and budgetary requirements. All of these situations require the gathering of information, modelling through mathematizing, drawing conclusions, and making predictions. While the teacher may provide some of the information and place restrictions on what the situation requires, the best opportunities are created when pupils work in cooperative groups to ask important questions, determine the type of information they require, and then work together to formulate a model.

Before discussing the importance of cooperative groups, it should be noted that if pupils are more familiar with decontextualised mathematics questions from textbooks, providing them with real-world problems might not be initially successful. Verschaffel and De Corte (1997) note that upper primary pupils frequently neglect real-world knowledge and realistic considerations when solving modelling tasks. In a teaching experiment to compare the use of realistic modelling with a regular curriculum, Verschaffel and De Corte noted pupils who experienced a range of realistic problems improved their disposition to solving real-world problems and were able to transfer their approaches to new situations. However, the use of modelling tasks in the teaching experiment was accompanied by a change in the variety of teaching methods used, and the establishment of new social and sociomathematical norms in the classroom. In particular, cooperative groups were used to "induce the intended cognitive conflicts and the subsequent reflections and cognitive changes" (Verschaffel & De Corte, 1997, p. 582). Through discussions with the whole class and feedback to groups, the teacher created a new classroom culture addressing pupils' beliefs about what counts as:

- a good mathematics problem;
- a good solution strategy:
- a good response; and
- a good explanation (p. 583).

3 The Importance of Cooperative Groups

In the world of work beyond school, the types of problems encountered are usually complex and multi-dimensional so solving problems requires teams of people working together to model situations and develop strategies and solutions. Each member of the team brings different knowledge and experiences to the group so they need to find ways to develop collegiality and collaborate to develop a plan and solve the problem. These skills are critical to successful problem solving and modelling and need to be developed throughout primary and secondary schooling (Anderson & White, 2004). To value all pupils unique contributions, Cohen et al. (1995) promote the notion of pupils having "multiple abilities" which recognises the different strengths pupils have which are required to solve problems.

Since modelling tasks usually require the gathering of information, and the need to pose questions and consider different perspectives and points of view, they are best attempted by small groups of pupils working towards a common goal. Clearly this takes time and many teachers feel they do not have extended periods of time to allow pupils the freedom to investigate such situations. However, given the dynamic nature of the modelling process and the different knowledge and understandings pupils will bring to the situation, working together can save time as well.

Pupils may suggest the possibility of different models but this is an important component of the process since it allows rich discussion about the advantages and disadvantages of the models proposed. Developing an argument for the use of a particular model is critical and provides opportunities for pupils to articulate their assumptions about the situation to be modelled. Pupils learn to listen to each other, to negotiate strategies and to support each other's efforts to understand and complete the task at hand. However, this does not always occur automatically and requires planning by the teacher.

Organising groups, allocating roles and responsibilities, and managing group efforts in the classroom are important considerations before implementing problem solving or modelling with primary school

pupils. For groups to successfully cooperate to solve modelling tasks, the groups need to be **structured** so that:
- positive interdependence exists among group members where all group members work together to complete the task;
- promotive interaction is encouraged with face-to-face discussions between pupils;
- individual accountability is required and accepted by all members of the group;
- interpersonal and small-group skills are employed with the teaching of skills as part of introducing cooperative groups into mathematics lessons; and
- group processing is practised so that members reflect on their work, and use metacognitive processes to monitor their progress (Gillies, 2007, pp. 4-5).

Ideally, groups should include three to four pupils with mixed knowledge, skills and understandings. All pupils need to be able to make a contribution to the solution of the task and they will each bring their unique attributes to benefit the group. Encouraging all pupils to take part, to listen to each other, and to value everyone's contribution may involve teacher modelling of this approach and teacher intervention while pupils work on the task. If the pupils are not used to working in cooperative groups, begin with pairs of pupils working together on shorter tasks and spend time identifying key 'rules' (Mercer, Wegerif & Dawes, 1999) for effective group work such as:
1. all information is shared;
2. the group seeks to reach agreement;
3. the group takes responsibility for decisions;
4. reasons are expected;
5. challenges to ideas are accepted;
6. alternative ideas are discussed; and
7. all group members are encouraged to speak to other members.

For younger pupils, the need to discuss taking turns and sharing tasks might be necessary while for older pupils, making decisions democratically, trying to understand another person's perspective, and providing constructive criticism would form part of the discussions of how to work in a cooperative group. While pupils are working together, teachers need to probe and clarify issues which arise, acknowledge and validate group efforts, confront discrepancies and clarify opinions, and offer tentative suggestions but only when absolutely necessary (Gillies, 2007). It is best not to provide answers since this encourages 'asking the teacher' and discourages group autonomy in decision-making.

In mathematics, there is evidence indicating problem solving can be enhanced when pupils work together in cooperative groups (e.g., Sahlberg & Berry, 2002; Terwel, 2003). However as Sahlberg and Berry (2002) note, this needs to be planned and well structured with pupils 'trained and guided' in appropriate cooperative behaviours. The authors also note that some tasks are more appropriate for cooperative work than others including open-ended investigations and modelling tasks which promote equal exchange between group members. For Sahlberg and Berry (2002, p. 89), modelling tasks are defined as "real problems that require mathematical principles and formulas in order to be solved. Tasks are open in terms of procedures and outcomes" with the following example provided:

You have a roll of kitchen paper and one sheet of identical paper. Estimate the number of sheets that your kitchen roll has in total (p. 88).

Task choice is critical for successful cooperative problem solving and modelling, particularly in mathematics (Anderson, 2005). If the task only requires pupils to exchange information and explanations or request assistance, this will lead to low levels of cooperation, lower levels of discussion and debate, and potentially lower levels of thinking. Rich modelling tasks are open with opportunity to model the real-life situation in a variety of ways so pupils need to share planning, decision making and often need to allocate tasks to achieve an outcome in a limited time.

4 Developing Modelling Tasks

As noted earlier, modelling tasks can be designed by investigating issues in school and local contexts, but they can also be developed by using data found in the local community, identifying and exploring stories from newspapers and magazines, considering issues raised in books or novels, or by asking mathematical questions based on photographs. Galbraith (2006) describes how ideas can be turned into modelling problems using a set of principles which include relevance and motivation, accessibility, feasibility of approach, feasibility of outcome, validity, and didactical flexibility where sequential questions retain the integrity of the real situation. So when developing tasks, begin by asking the following questions:

1. What are the interests and everyday experiences of the pupils which could be used to engage the pupils in the task? (e.g., sport, recreational games or puzzles, travel);
2. What sources of information could be used and how accessible are they to the pupils? (e.g., internet, travel magazines);
3. How much time is required to start and then carry out the task? (how many mathematics lessons will be required);
4. What support will be required to assist pupils begin the task? (consider assumptions, pose new questions).
5. Do the pupils need to do "readiness questions" (English, 2008) to ensure they all understand the context and the information provided?

5 Examples of Modelling Tasks

The following examples provide starting points for possible investigations or modelling tasks. It is recommended that teachers choose appropriate tasks which their pupils will find interesting and engaging. Alternatively encourage pupils to pose their own questions based on stimulus material.

5.1 *Investigating the world's people*

Use the book *If the World Were a Village of 100 People* by Smith (2006) to stimulate student investigations of the world they live in. In the book Smith presents information about the world's population by representing the population of approximately 6.4 billion people as just 100 people in the village. That means that each person in the village represents 6 million 400 thousand people. From this he presents data about the village so that pupils get a sense of the proportions of the population represented by particular groups and become aware of important issues and inequities in society – for example, 22 people speak a Chinese dialect, 17 people cannot read or write, 50 people are often hungry and only 24 people have a television in their home.

By designing a model to represent the people in the world as well as in countries such as Singapore or Australia, pupils could compare the data to determine if any country might reflect similar characteristics to the world's population and then consider why this might or might not be possible. This modelling task integrates several areas of the curriculum with the potential to explore concepts in mathematics, geography, science, and history as well as to investigate current affairs and global issues. Additional suggestions for investigations using this book are presented in Figure 3 and Figure 4.

David J. Smith, the author of *If the World Were a Village*, has been asked to write a sequel to the original book. This time, he would like to base his data on a classroom. The task is to research whether a single classroom could represent the world and to write a letter reporting the findings.

Small groups could collect data from pupils in the class and determine how to display the information. The data need to be compared to the original book as well as other classes.

Figure 3. Does our classroom represent the world?

The editor of a travel magazine wants to include an article about the best countries in which to live.

Your job is to present information that compares the standard of living in Singapore with that of three other countries on three different continents.

Use a variety of methods including data displays, data analyses and maps to support your conclusion.

Figure 4. What are the best countries to live in?

5.2 Investigating sporting contexts

The Olympic Games events receive world-wide media coverage. If pupils are interested in sporting events and like to follow particular athletes or performances of particular countries, the medal table raises questions about how countries should be compared.

1. Is it fair to determine the overall winner on the basis of the highest number of gold medals?
2. What difference would it make to the results if the total number of medals were considered?
3. Are there other ways to examine the medal tally table to compare countries?
4. What factors should be taken into consideration when comparing countries?

Pupils could design models including some of the identified factors and revise the table to justify which country should be declared the winner. Comparing different models would yield potentially different results and enable considerable debate about the best approach. Table 1 presents the medal tallies for some of the countries from the Beijing Olympic Games in 2008. These could be used as a starting point for the investigation although pupils may want to consider many more countries or particular types of countries.

Table 1

Final medal tally for the Summer Olympic Games, Beijing, China, 2008

Country	Gold	Silver	Bronze
China	51	21	28
USA	36	38	36
Russia	23	21	28
Great Britain	19	13	15
Germany	16	10	15
Australia	14	15	17
South Korea	13	10	8
Japan	9	6	10
Indonesia	1	1	3
Singapore	0	1	0

English (2007a) developed a modelling task for Australian pupils which required them to determine the composition of the Australian swimming team for the 2004 Olympic Games and for the 2006 Commonwealth Games. Pupils from fourth-grade were given background information and 'readiness questions' to ensure understanding of the context and the complex information provided. In addition, they were given a set of data about swimmer's most recent times and were required to recommend the two best swimmers to the selection committee. A letter had to be written with the recommendation and needed to include the reasons for their choice of team members. This task required considerably more information than the one provided above but was highly successful with pupils using a diversity of models to solve the task. English (2007a, p. 7) notes the importance of these experiences if pupils are to be prepared to deal "with our increasingly complex, data-drenched world".

5.3 *Investigating information from the media*

Data are frequently presented in newspapers or magazines in may different forms. These may be used as a stimulus to pose questions for investigation. The information presented in Table 2 was originally published in a newspaper in England in 2006. To begin the investigation, pupils could be asked to pose their own questions or the following could be used as a starting point.
1. What questions could be asked about the data?
2. How would the information have been collected to compile these percentages?
3. What changes have occurred in lifestyle between 1957 and 2006?
4. How would this impact on household spending and why?
5. If similar data were collected in Singapore, what differences might there be between the data for England and the data for Singapore?

Table 2
How spending has changed in England – percentage of household expenditure

Items	1957	2006
Housing	9	19
Fuel and Power	6	3
Food	33	15
Alcohol	3	3
Tobacco	6	1
Clothes and Shoes	10	5
Household goods	8	8
Other goods and services	16	29
Transport and vehicles	8	16

5.4 *Investigating other contexts*

Researchers have used many other contexts to explore the potential of particular modelling tasks to engage pupils in real-life investigations using significant mathematical ideas. For example, English, Fox and Watters (2005) describe an investigation of cyclone disasters in northern Australia, a situation particularly relevant to the pupils in Queensland who may experience such disasters in summer months. Pupils were presented with considerable information about recent cyclones and were required in their investigation to recommend two areas in Queensland which should be avoided in the location of a new holiday resort. Given the extent of the information provided, pupils were required to make critical decisions to:

- distinguish relevant from irrelevant information;
- determine the most important factors for consideration (e.g., flooding, wind speed, power loss, deaths, …);
- develop an appropriate model including which mathematical operations to apply to the data; and
- represent the findings in the most appropriate way.

English and her colleagues have published widely with a range of suitable tasks (e.g., English, 2007b, 2008; English, Fox, & Watters,

2005; English & Watters, 2004). Other worthwhile examples can be located in Chamberlin and Zawojewski (2006) and Swetz and Hartzler (1991).

6 The Potential of Modelling to Develop Pupils' Problem Solving Competencies

While problem solving has been recognised as an important component of the school mathematics curriculum, attempts to incorporate problem solving into mathematics lessons have not been as successful as we would like. There are many reasons for this:
- teachers' beliefs about the purpose and role of problem solving in learning mathematics;
- teachers' confidence about using problem solving approaches to teaching mathematics;
- time to prepare and plan good problem-solving tasks which are suitable for all pupils;
- too much content in the curriculum; and
- assessment practices which focus on lower-order thinking (Anderson, 2009).

Using modelling tasks is one possible solution to supporting teachers' efforts to implement problem solving in a way which engages pupils and enables the development of deeper mathematical understanding as well as a range of problem-solving skills and competencies. Mousoulides, Sriraman and Christou (2007) argue for a future oriented approach to mathematical problem solving using a models and modelling approach. If teachers are to provide opportunities for pupils to develop all of the skills and competencies identified by Stacey (2005) and presented in Figure 1, using rich modelling tasks and small group cooperative learning is a potential answer to the dilemma of finding ways to implement the problem-solving requirements of the mathematics curriculum.

References

Anderson, J. (2005). Working mathematically in NSW classrooms: An opportunity to implement quality teaching and learning. In M. Coupland, J. Anderson, & T. Spencer (Eds.), *Making Mathematics Vital* (pp. 53-59), Proceedings of the 20th biennial conference of the Australian Association of Mathematics Teachers. Sydney: AAMT.

Anderson, J. (2009). *Mathematics Curriculum Development and the Role of Problem Solving*. Paper presented at the biennial conference of the Australian Curriculum Studies Association – to be published on the ACSA website.

Anderson, J., & White, P. (2004). Problem solving in learning and teaching mathematics. In B. Perry, G. Anthony, & C. Diezmann (Eds.), *Research in mathematics education in Australasia 2000-2003* (pp. 127-150). Flaxton, Qld: Post Pressed.

Chamberlin, M. T., & Zawojewski, J. S. (2006). A worthwhile mathematical task for students and their teachers. *Mathematics Teaching in the Middle School*, *12*(2), 82-87.

Cohen, E. G., Lotan, R. A., Whitcomb, J. A., Balderrama, M. V., Cossey, R., & Swanson, P. E. (1995). Complex instruction: Higher-order thinking in heterogeneous classrooms. In R. J. Stahl (Ed.), *Handbook of Cooperative Learning Methods* (pp. 82-96). Westport, CT: Greenwood Press.

English, L. (2007a). *Modeling with complex data in the primary school*. Paper presented at the 13th biennial conference of the International Community of Teachers of Mathematical Modelling and Applications, Indiana University, 22 to 26 July.

English, L. (2007b). Interdisciplinary modelling in the primary mathematics classroom. In J. Watson & K. Beswick (Eds.), *Mathematics: Essential research, essential practice*, (Proceedings of the 30th annual conference of the Mathematics Education Research Group of Australasia, pp. 275-284). Hobart: MERGA.

English, L. D. (2008). Introducing complex systems into the mathematics curriculum. *Teaching Children Mathematics*, *15*(1), 38-45.

English, L. D., Fox, J. L., Watters, J. J. (2005). Problem posing and solving with mathematical modelling. *Teaching Children Mathematics*, *12*(3), 156-163.

English, L., & Watters, J. (2004). Mathematical modelling in the early school years. *Mathematics Education Research Journal*, *16*(3), 59-80.

Galbraith, P. (2006). Real world problems: Developing principles of design. In In P. Grootenboer, R. Zevenbergen, & M. Chinnappan (Eds.), *Identities, cultures and learning spaces, Volume 1*, (pp. 229-236). Proceedings of the 29th Annual Conference of the Mathematics Education Research Group of Australasia, Canberra, ACT.

Gillies, R. M. (2007). *Cooperative learning: Integrating theory and practice.* Los Angeles: Sage Publications

Lesh, R., Hoover, M., Hole, B., Kelly, A., & Post, T. (2000). Principles for developing thought-revealing activities for students and teachers. In A. E. Kelly & R. A Lesh (Eds.), *Handbook of Research Design in Mathematics and Science Education* (pp. 591-646), Mahwah, NJ: Lawrence Erlbaum Associates.

Mercer, N., Wegerif, R., & Dawes, L. (1999). Children's talk and the development of reasoning in the classroom. *British Educational Research Journal, 25*, 95-111.

Ministry of Education (2006). *Mathematics syllabus: Primary.* Singapore: Author.

Mousoulides, N. G., Sriraman, B., & Christou, C. (2007). From problem solving to modelling – the emergence of models and modelling perspectives. *Nordic Studies in Mathematics Education, 12*(1), 23-47.

National Curriculum Board (2009). *Shape of the Australian curriculum: Mathematics.* Barton, ACT: NCB.

Polya, G. (1971). *How to solve it.* Revised edition. Princeton University Press.

Sahlberg, P., & Berry, J. (2002). One and one is sometimes three in small group mathematics learning. *Asia Pacific Journal of Education, 22*(1), 82-94.

Smith, D. J. (2006). *If the world were a village of one hundred people: A book about the world's people.* Sydney: Allen and Unwin.

Stacey, K. (2005). The place of problem solving in contemporary mathematics curriculum documents. *Journal of Mathematical Behaviour, 24*, 341–350.

Swetz, F., & Hartzler, J. S. (1991). *Mathematical modeling in the secondary school curriculum.* Reston, VA: National Council of Teachers of Mathematics.

Terwel, J. (2003). Co-operative learning in secondary education: A curriculum perspective. In R. M. Gillies & A. F. Ashman (Eds.), *Co-operative Learning: The Social and Intellectual Outcomes of Learning in Groups* (pp. 54-68). London: RoutledgeFalmer.

Verschaffel, L., & De Corte, E. (1997). Teaching realistic mathematical modeling in the elementary school: A teaching experiment with fifth graders. *Journal for Research in Mathematics Education, 28*(5), 577-601.

Zawojewski, J., Lesh, R., & English, L. (2003). A models and modeling perspective on the role of small group learning activities. In R. Lesh & H. M. Doerr (Eds.), *Beyond constructivism: Models and modeling perspectives on mathematics problem solving, learning, and teaching* (pp. 337-358). Mahwah, NJ: Lawrence Erlbaum.

Zawojewski, J. (2007). *Problem solving versus modelling.* Paper presented at the 13th biennial conference of the International Community of Teachers of Mathematical Modelling and Applications, Indiana University, 22 to 26 July.

Chapter 6

Word Problems and Modelling in Primary School Mathematics

Jaguthsing DINDYAL

Word problems are generally used for teaching about the applications of mathematics and they provide an excellent avenue for primary students to engage in modelling activities. This chapter focuses on word problems from a modelling perspective and provides a categorisation of such problems that teachers will find helpful in their teaching. Some issues about the use of word problems as modelling tasks and some hints for teaching are also highlighted.

1 Introduction

It is well-documented that students need to apply the mathematics they learn in school. Application of mathematics involves the use of mathematical tools to analyse some aspect of a real-life or realistic situation. The teaching about the applications of mathematics should start early, as early as we start the teaching of mathematics itself. The application of mathematics cannot be divorced from the use of models and the modelling process. As most application problems involve the discussion of some reality or "realistic contexts", the use of models to access that reality becomes essential. The models, thus, mediate between the realistic context and mathematics, and they help in translating the essential relationships into mathematics for a purely mathematical analysis. Accordingly, the use of models and the modelling process are important aspects in the learning of mathematics.

At the primary level, one way that students can apply the mathematics they have learned is through word problems that involve modelling. While teachers are familiar with word problems the use of modelling is not so familiar. Thus this chapter starts by addressing the following questions: What is modelling? Why include modelling in school mathematics? How can modelling be brought to the classroom? What issues are related to modelling? First, I focus on the terms models and modelling in the context of the primary level mathematics. Then, I highlight word problems from the perspective of modelling and provide a categorisation of such problems for teaching at this level.

2 Models

A model can be described as something that one manipulates to find out about something else (see Mason & Davis, 1991). A large object like an aeroplane can be studied by drawing a scaled-down model or a very small object like an atom can be studied by drawing a scaled-up model. Models help us to contract and expand time and space and to manipulate and experiment without disturbing the actual situation. Thus, a model can be manipulated and studied and in the process, information on the parent object can be obtained (Swetz & Hartzler, 1991). The models need not always be physical objects; they can also be visual sketches, schemes, diagrams and even symbols (Van Den Heuvel-Panhuizen, 2003). We can also consider models of mathematical concepts as described by Van de Walle (2004):

> A model for a mathematical concept refers to any object, picture, or drawing that represents the concept or onto which the relationship for that concept can be imposed" (p. 28).

In application problems, children very often have to deal with problems in realistic contexts which can only be made accessible and *manipulable* if models are used. Thus, models have an important role in making mathematics become real for students. It should be noted that a model is not something universal. It is always from the perspective of the modeller. Verschaffel (2002) claimed that:

In modeling, not all aspects of reality can, nor should, be modeled. The modeler tries to capture the essentials of the situation in the model, but what is considered as essential, and, thus, taken into account in the mathematical model, is not fixed, but is relative to the modeler and to the context wherein (s)he is confronted with the modeling task at hand (p. 64).

As such, the same reality can be modelled in quite different ways by different individuals. The implication is that students who are involved in modelling activities may come up with completely different conceptualisations of problem situations as their models may be quite different from each other.

3 Modelling

Modelling is involved when there is the need to create a model to capture the essence of some piece of reality and subsequently, to manipulate the model to arrive at some conclusions about the piece of reality. Swetz and Hartzler (1991) described mathematical modelling as a systematic process that draws on many skills and employs the higher cognitive activities of interpretation, analysis, and synthesis. The authors suggested that the modelling process is composed of four main stages (see Figure 1):

1. Observing a phenomenon, delineating the problem situation inherent in the phenomenon, and discerning the important factors (variables/parameters) that affect the problem.
2. Conjecturing the relationships among factors, and interpreting them mathematically to obtain a model of the phenomenon.
3. Applying appropriate mathematical analysis to the model.
4. Obtaining results and reinterpreting them in the context of the phenomenon under study and drawing conclusions. (p. 2)

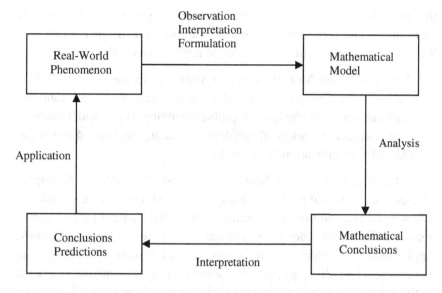

Figure 1. The Modelling Cycle

The authors also suggested that a fifth stage to this modelling process could be the testing and refining of the model. As can be noted in Figure 1, the modelling process involves the development of a model. However, the model is not an end in itself. The model serves to bridge the reality described in the problem text and the world of mathematics. The modelling process, in addition, includes the mathematical analysis to reach some conclusions and a reinterpretation of the solution in the realistic context of the problem situation. Other similar approaches to the modelling process have been suggested (see Mason & Davis, 1991; Niss, 1989). The modelling cycle described above in Figure 1 may need some adaptation for students at the primary level (see Figure 2).

3.1 Why include modelling at the primary level?

Opportunities to explore real-life applications make mathematics more meaningful for students and aid in the development of important other skills. Since applications involve the use of models and some aspects of the modelling process, modelling cannot be ignored in the primary curriculum. The curriculum document from the Ministry of Education in Singapore emphasises the importance of applications and modelling:

> Application and Modelling play a vital role in the development of mathematical understanding and competencies. It is important that students apply mathematical problem-solving skills and reasoning skills to tackle a variety of problems, including real-world problems. (Ministry of Education, 2006, p. 8)

The emphasis on application and modelling in the Singapore Mathematics Curriculum Framework (SMCF) should not be trivialised. It is not just the addition of something new in the curriculum and should not as well be considered as a topic area to be studied separately. Application and modelling cut across various content domains within mathematics and although they may seem new, they have always existed in the school curriculum. However, the emphasis may not have been quite the same. The *Model Method* has been used in the Singapore mathematics curriculum at primary level since the 1990s (for a detailed description of the Model Method, see Curriculum Planning and Development Division, 2009; Kho, 1987). However, the Model Method should not be equated with modelling, although the use of this method involves an aspect of modelling. The renewed interest in modelling probably stems from corresponding changes taking place in mathematics curricula all over the world. Mathematics educators wish to make mathematics more meaningful to students by highlighting the importance of mathematics in solving real-world problems.

4 Word Problems

Word problems (also called story problems or verbal problems) have a long history in mathematics teaching, especially at the primary level. Verschaffel, Greer and De Corte (2000) describe word problems as verbal descriptions of problem situations, typically presented in a school context, wherein one or more questions are raised, the answer to which can be obtained by the application of mathematical operations to numerical data available in the problem statement. These problems are generally used by teachers to test for higher order thinking skills and are usually perceived as being harder by the students. Word problems describe some kind of reality or a realistic context for solving problems. It is understood that the realistic aspect is from the perspective of the student who will be engaged in solving the problem. Consider the following problem:

John has 24 stickers. Mary has twice as many. How many stickers do they have altogether?

The simple word problem given above describes a context or some type of reality. There are some key words such as "twice" and "altogether". A child who is solving this problem needs to accept the problem as one to be solved and use a mathematical model to solve the problem. Although the context seems to categorise this problem as an arithmetic problem, it is the method used to solve the problem that actually decides the categorisation. This problem can also be solved using algebra, if one wishes to do so.

Why do we need word problems? Judiciously chosen word problems should be an integral part of mathematics teaching at the primary level.

Also, we should be careful about how and when to use them in class. For children to have a thorough knowledge of the mathematics they have learned, they must be able to apply the mathematics in realistic situations. Some points to note include:
1. Word problems provide an excellent opportunity to children for applying the mathematics that they have learned. They also provide children with a basic sense and experience in mathematization, especially mathematical modelling (Reusser & Stebler, 1997).
2. Carefully selected word problems can make real-life or realistic situations accessible to students for them to apply the mathematics that they have learnt.
3. Rather than apply mathematical operations in a purely mathematical domain, word problems provide a context for doing so. Using

mathematics within a particular context is an important life-skill that we wish children to cultivate. Unfortunately, quite often children overlook contextual information in a problem; they focus on some key words; and do not interpret their solution in the context highlighted. Consider the example: *A car can carry 5 students. How many cars are needed to carry 42 students to a museum?* A student who overlooks the contextual information may give 8, 8.4 or 8 remainder 2, as an answer.
4. Real-life or realistic contexts in well-crafted word problems demonstrate the *messiness* of such problems when compared to the traditional problems involving usually *nice* numbers in school textbooks. Such problems may involve extraneous information that students will have to learn how to cope with when solving the problems.

4.1 Solving word problems

Word problems or story problems involve some type of application of mathematics, and as such, some aspect of modelling is involved. The modelling cycle described in Figure 1 is relevant for solving general modelling tasks, however for students at the primary level, the modelling process described by Verschaffel, Greer and De Corte (2000) seems more appropriate. These authors described the following phases in the modelling process (see Figure 2):
1. understanding the situation described;
2. constructing a mathematical model that describes the essence of those elements and relations embedded in the situation that are relevant;
3. working through the mathematical model to identify what follows from it;
4. interpreting the outcome of the computational work to arrive at a practical situation that gave rise to the model;
5. evaluating that interpreted outcome in relation to the original situation; and
6. communicating the interpreted results. (p. xii)

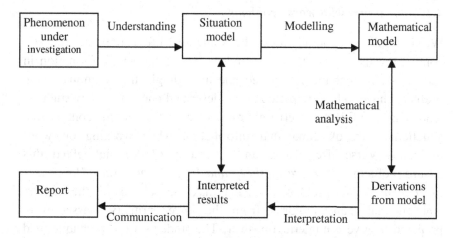

Figure 2. The process of modelling
Source: Verschaffel, Greer and De Corte (2000, p. xii)

Using this process, the above authors claim that even the simplest word problem can be viewed as a modelling exercise. Consider the simple problem: *Tom has 8 marbles. 5 of the marbles are red. How many are not red?* (see Figure 3)

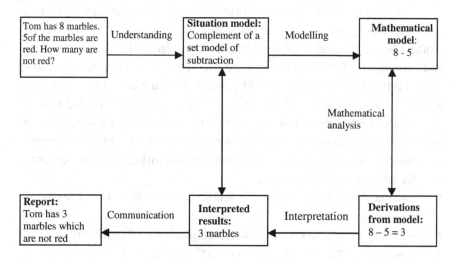

Figure 3. A simple word problem as a modelling exercise

4.2 Some issues with word problems

Word problems generally describe some real-life context or realistic context that is accessible to the student. The contextual information in such problems cannot be ignored and accordingly the solutions to the problems have to be interpreted in the described contexts. Many students who can confidently perform routine calculations in context-free situations overlook contextual information when working on word problems. Verschaffel, Greer and De Corte (2000) highlighted this absurd problem: *There are 26 sheep and 10 goats on a ship. How old is the captain?* This problem does not make sense and yet the authors claimed that many children from the lower primary classes were prepared to give a numerical answer. The students were probably cued by the numbers in the problem situation that some operations were involved and that the context did not really matter. The emphasis was on getting an answer at all costs. By overlooking the contextual information these students tended to move directly from the textual information to the mathematical model, to the derivations from the mathematical model and then to the reporting (see Figure 3). As such, many children bypass critically important stages in the solving process.

Verschaffel, Greer and De Corte (2000) pointed out that there is a strong tendency on the part of many children to exclude relevant real-world knowledge and realistic considerations from their understanding and solving of school arithmetic word problems. It is important to know why children ignore the real-world context. To what extent is this due to ignorance of the real-world context or to a misapplication on the part of the children? Reusser and Stebler (1997, pp. 324-325) who worked with lower and upper secondary students, identified some contextual rules and assumptions on the part of students that seem to influence their decision, to solve mathematical problems:

- Assume that every problem presented by a teacher or in a textbook make sense.
- Do not question the correctness or completeness of problems.
- Assume that there is only one "correct" answer to every problem.
- Give an answer to every problem presented to you.

- Use all numbers that are part of the problem in order to calculate the solution.
- If a chosen mathematical operation works out without remainder (evenly), you are probably on the right track.
- If a problem is perceived to be indeterminate, equivocal, or unsolvable, go for an obvious interpretation given the information in the problem text and your knowledge of mathematical operations.
- If you do not understand a problem, look at key words, or at previously solved problems in order to determine a mathematical operation.

Reusser and Stebler also highlighted that:

Ample evidence shows a clear tendency of children to neglect realistic considerations and to exclude real-world knowledge from their mathematical problem solving" (p. 310).

Lets us look at some word problems whereby the problem context poses a constraint or is problematic for a student who overlooks the context. For example:

Mrs Tan spends $400 in 45 minutes in a big store. How much will she spend in 3 hours?

The problem involves understanding a realistic context. In this context, proportionality cannot be assumed for the amount spent in a given period of time.

A further example is as follows:

Mellissa jogs for 2 hours on Sunday. How many hours does she jog in a week?

This is not a simple proportional reasoning problem. The contextual information is very important. Does Melissa jog for 2 hours every day of the week? The relevant question here is whether we should be using such problems which have serious realistic constraints. Real-life problems or problems with realistic contexts are quite often problematic as described above. If we wish children to solve real-world problems and improve their modelling skills, it is imperative that they be exposed to such problems. Unfortunately, most textbook problems tend to have nice numbers within contexts described ideally to produce nice solutions. Thus, we note that the use of word problems in the mathematics classroom is not unproblematic. The point to ponder about for the teacher is: To what extent can we say that the overlooking of contextual information by students is due to an inherent issue in the design of the problem or due to a lack of knowledge about the contextual information on the part of the student?

4.3 *Classification of word problems from a modelling perspective*

It is obvious that modelling problems have to be as authentic as possible. To what extent is a problem authentic? Galbraith and Stillman (2001, pp. 301-302) have proposed a classification of word problems from a modelling perspective:

Injudicious problems. These are problems which are somewhat detached from reality. For example:

> *One person can dig a hole in the ground in 6 hours. 10 persons working at the same rate can dig the hole in how much time?*

Although, it is possible to find a numerical answer to this problem, it does not make much sense for 10 persons to be digging the same hole at the same rate and at the same time.

Context-separable problems. In these problems the context is quite artificial and can be stripped away to expose a purely mathematical problem. For example:

> *Kevin goes to the shop with 10 coins, which amongst others include some 10-cent, 20-cent, and 50-cent coins. What is the largest amount of money that he can spend in the shop?*

The context of going to the shop is not really important and can be ignored. We can simply ask for the largest amount of money that can be spent.

Standard application problems. In these problems the mathematics is context-related and the situation is realistic. However, the procedure is standard as the solver is cued to some essential information. For example:

> *One bus can sit 30 passengers. Find the number of buses required to take 250 persons to a shopping mall (no person can be left out).*

There is a cue that no person can be left out. Quite often the reason why such problems are set is that we wish children to understand that nobody can be left out.

Modelling problems. These are typical modelling problems in which no mathematics appears in the problem statement, where the formulation of the problem, in mathematical terms, must be supplied by the modeller. For example:

> *Tom wishes to find out how many apples are eaten by the students in his school in a month. Explain how he can do so.*

This problem is quite different from the other problems and will be more demanding on the average child as there is little support provided. Some kind of data will have to be collected and analysed. The modelling cycle suggested by Swetz and Hartzler (1991) can be used here.

We certainly want to avoid the injudicious problems for children at the primary level, as we wish to provide realistic contexts for children to make sense of the mathematics they learn in school. However, the other types of problems are still valuable. The modelling problems which look more authentic, if accompanied with appropriate teacher guidance and support may be successful with some students (English & Walters, 2004; English, 2007). However, we need to be careful as not all such problems may be accessible to children at the primary level.

Modelling problems at the primary level can also be classified according to the degree to which some structure is provided in the problem. From this perspective, a broad classification of word problems may include three types of problems: structured problems, semi-structured problems and unstructured problems. Description and an example of each follows:

Structured problems: Structured problems are the traditional word problems in which a plausible real context is described and all necessary information is provided. The students do not have to collect any data or make any type of measurements. They already know what are the numbers involved. The problem is closed as the data provided in the problem leave little room creative responses.

Tom has 24 marbles. Jerry has 18 marbles. How many more marbles than Tom does Jerry have?

We note that all of the necessary information for solving the problem is provided. The students have to make sense of the context and understand the meaning of some key words such as "How many more..." Problems described by Galbraith and Stillman (2001) as being context-separable or standard modelling problems and even the injudicious problems can be included in this category.

Semi-structured Problems: In this type of problems, a real-life context is described and some data are usually provided, generally in a table. Students have to interpret the given data to complete the modelling task. The questions are open-ended and students have the opportunity to provide some creative responses (see Figure 4).

Unstructured Problems: This type of problems is called as typical modelling problems by Galbraith and Stillman (2001). These are modelling problems in which no mathematics appears in the problem statement, where the formulation of the problem, in mathematical terms, must be supplied by the modeller.

Find the total number of coins that students in your school carry on a particular day.

	1970	1980	1990	2000	2007
0-14 years	39.1	27.6	23.0	21.9	18.9
15-64 years	57.5	67.5	71.0	70.9	72.5
65 years and over	3.4	4.9	6.0	7.2	8.5
Age composition (%)	100	100	100	100	100
Total Population (thousands)	2074.5	2413.9	3047.1	4027.9	4588.6

The table above shows how the age composition of the population in Singapore has changed over time from 1970 to 2007.

Source: www.singstat.gov.sg

(i) How many persons aged 15-64 years old lived in Singapore in 1990?

(ii) How many persons aged 65 years or more lived in Singapore in 2000?

(iii) Write a report describing how the age composition of the population has changed over time. What do you predict as the age composition of the population for 2010? For 2015? Explain your reasoning.

Figure 4. A semi-structured problem

This problem is quite different from the other problems and will be more demanding on the average child as there is little support or structure provided for solving the problem. Some estimation need to be made and some data need to be collected. Students can model such problems in various ways and can provide creative responses.

4.4 *Notes for the teacher when using word problems as modelling tasks*

Using effectively in classroom instruction, word problems can provide children with opportunities to apply the mathematics that they have learned. However, unless word problems are crafted carefully, they run the risk of being stereotypical and thus encouraging children to develop certain standard solution procedures. Teachers should plan to use all three types of problems described above. If students are not accustomed to modelling activities then some pre-modelling activities can help to better prepare them for the modelling tasks. Pre-modelling activities that the teachers may use before embarking on modelling tasks that are

typically semi-structured or unstructured include doing the following (see Dindyal, 2009):
1. Dispel the myth of a single correct solution for a given problem. Prior to solving typical modelling problems, give some practice with open-ended problems as most modelling problems do not have single correct solutions.
2. Use authentic data from reliable sources to construct problems. Modelling problems involve data from realistic contexts and are usually not nice rounded numbers as portrayed in typical textbook problems.
3. Provide ample estimation activities prior to giving modelling problems to students. "Estimation begins with guessing, but as the guessing becomes more and more informed, it becomes like a model" (Mason & Davis, 1991, p. 21). As such estimation involves implicit modelling.
4. Set some problems with realistic constraints, such as: *Tim counts 10 cars in 5 minutes while walking home from school. How many cars will he count in the next 10 minutes?* This is not a simple proportional reasoning problem. The context imposes some constraints. Modelling problems invariably involve such realistic constraints which cannot be ignored by the teacher who wishes to cultivate good modelling skills in his or her students.

Word problems that would be used as modelling tasks have to be carefully planned by the teacher. Although teachers cannot predict beforehand the kinds of models and modelling that students will use, it is advisable that teachers anticipate the modelling of those particular situations themselves. Teachers who carry out the modelling exercise themselves are more confident when responding to students' questions.

Some other points to note by teachers about word problems used as modelling tasks include the following (see Dindyal, 2009):
1. A context that is unfamiliar can become a major hurdle for a child who wants to model the situation described in the problem. Accordingly, it is important to start with modelling situations that are not too complex for the child.
2. Word problems are heavily dependent on the language used in the text to describe the reality or the realistic context. The semantics can

become a major obstacle for children getting a good grasp of the reality described in the problem text. This may turn off children with poor language background or paint a different picture for some children.
3. In the solution process, it is important that the teacher does not direct the students to the type of mathematics that they will be using while modelling the problem situation. It is better to let students use mathematical procedures that they are more confident to use.
4. If word problems of a similar nature are repeated for the same group of children then we run the risk of *routinising* the modelling procedure which defeats the purpose of using the word problems in the first place. These problems can become disguised practice problems.
5. Too many word problems can lead to a general apprehension on the part of even very highly motivated children. Also, if the problems used in class do not match the ability level of the children, then they may get uninterested and we may run the risk of trivialising the modelling and application perspective that we wish to highlight for them.
6. Since modelling activities invariably involve some kind of report, it is advisable to inform students about the potential audience of the report. Students should be advised to write a statement of some sort at the very end of the solution process.

5 Conclusion

Unstructured modelling problems are certainly the most demanding type of activity as they involve some type of data collection and analysis. Such problems may not be within the reach of all pupils, although there may be some exceptions. It is better for teachers to start with the structured modelling problems and then proceed to the semi-structured modelling problems eventually leading to the unstructured modelling problems for those students who are more able.

To a purist, modelling is esoteric knowledge reserved for elites with very high level of competency in mathematics. However, modelling has a role in the primary curriculum. The lense to view modelling at the primary level has to be readjusted to match the mathematical sophistication and general education level of the learners. Word problems provide good avenues for making some type of realistic contexts accessible to our students at the primary level. However, the problems have to be judiciously chosen by the teacher and the students need to be adequately prepared to see the value of solving these types of problems.

References

Curriculum Planning and Development Division. (2009). *The Singapore model method for learning mathematics*. Singapore: Ministry of Education.
Dindyal, J. (2009). *Applications and modelling for the primary mathematics classroom*. Singapore: Pearson.
English, L. D. (2007). Complex systems in the elementary and middle school mathematics curriculum: A focus on modeling. *The Montana Mathematics Enthusiast, Monograph 3*, 139-156.
English, L., & Watters, J. (2004). Mathematical modelling with young children. In M. Hoines & A. Fuglestad (Eds.), *Proceedings of the 28th conference of the International Group for the Psychology of Mathematics Education* (Vol. 2, pp. 335-342). Bergen: PME.
Galbraith, P., & Stillman, G. (2001). Assumptions and context: Pursuing their role in modelling activity. In J. F. Matos, W. Blum, S. K. Houston, & S. P. Carreira (Eds.), *Modelling and mathematics education. ICTMA 9: Applications in science and technology* (pp. 300-310). Chichester, U.K.: Horwood.
Kho, T. H. (1987). Mathematical models for solving arithmetic problems. *Proceedings of the 4th Southeast Asian Conference on Mathematics Education* (ICMI-SEAMS). (pp. 345-351), Singapore.
Mason, J., & Davis, J. (1991). *Modelling with mathematics in primary and secondary schools*. Geelong, Vic, Australia: Deakin University Press.
Ministry of Education. (2006). *Mathematics Syllabus: Primary*. Singapore: Author.
Niss, M. (1989). Aims and scope of applications and modelling in mathematics curricula. In W. Blum, J. S. Berry, R. Biehler, I. Huntley, G. Kaiser-Messmer, & L. Profke

(Eds.), *Applications and modelling in learning and teaching mathematics* (pp. 22-31). New York: Ellis Horwood Limited.

Reusser, K., & Stebler, R. (1997). Every word problem has a solution – The social rationality of mathematical modelling in schools. *Learning and Instruction, 7*(4), 309-327.

Swetz, F., & Hartzler, J. S. (Eds.). (1991). *Mathematical modeling in the secondary classroom*. Reston, VA: National Council of Teachers of Mathematics.

Van de Walle, J. A. (2004). *Elementary and middle school mathematics: Teaching developmentally* (5th ed.). Boston: Pearson.

Van Den Heuvel-Panhuizen, M. (2003). The didactical use of models in realistic mathematics education: An example from a longitudinal trajectory on percentage. *Educational Studies in Mathematics, 54*, 9-35.

Verschaffel, L. (2002). *Taking the modeling perspective seriously at the elementary school level: Promises and pitfalls*. In Proceedings of the 26th Annual Meeting of the International Group for the Psychology of Mathematics Education (Vol. 1, pp. 64-80), Norwich, England, July 21-26 (Eric Document Reproduction Service No. ED 476 082).

Verschaffel, L., Greer, B., & De Corte, E. (2000). *Making sense of word problems*. Lisse, The Netherlands: Swets & Zeitlinger Publishers.

Chapter 7

Mathematical Modelling in a PBL Setting for Pupils: Features and Task Design

CHAN Chun Ming Eric

Contemporary pedagogies that are grounded in theory are making their way into mathematics classrooms. This is consistent with reformed efforts to empower learners to solve problems in a dynamic world. A problem-based learning (PBL) instructional approach is one such pedagogy. For teachers to enact PBL lessons they must know what PBL entails and be able to use suitable problem tasks. This knowledge is important to differentiate problem-solving from the traditional form of solving problems. Mathematical modelling tasks that stem from a modelling perspective are deemed as appropriate and befit the type of tasks used in a PBL setting. This chapter explains mathematical modelling, problem-solving in the context of a short cycle PBL instructional setting, and discusses the designing of mathematical modelling tasks. Well-designed tasks can enable pupils to manifest their mathematical thinking as they create and interpret situations.

1 Introduction

Reformed efforts are increasingly challenging the way pupils solve problems in mathematics classrooms. The preoccupation with the teacher demonstrating a series of steps and then having the pupils follow and practice solving similar word problems is deemed to limit the potential that pupils can demonstrate in solving problems. Not that solving structured word problems is unimportant as it should rightly have a place in the curriculum, but its over-emphasis can be unhelpful to prepare pupils in using mathematics beyond the classroom or solving problems in complex and dynamic situations. Many word problems present simplified forms of decontexualized real-world situations that are contrived and carried out only in the confines of a classroom. Some researchers regard solving traditional word problems as inadequate to address the mathematical knowledge, processes, representational fluency, and social skills that pupils need to develop for the 21st century (English, 2002; Lesh & Lehrer, 2003) and lament their lack of realistic constraints (Verschaffel, De Corte, & Borghart, 1997). Over time, solving such problems become mechanistic and behaviouristic (Chan, 2008).

Reformed efforts are looking towards learning environments where pupils can develop deeper understanding and apply higher-order thinking in mathematics learning. A problem-based learning (PBL) instructional approach is one such learning platform that is worth considering. One of the important distinguishing features of PBL is its emphasis on the task that is said to drive the learning. If PBL is to be implemented in schools, then it is critical that the tasks befit the characteristics of the type of tasks used in PBL settings. Modelling tasks designed from a models-and-modelling perspective (Lesh & Doerr, 2003) are deemed to be appropriate for use in PBL settings (Hjalmarson & Diefes-Dux, 2007). When pupils engage in mathematical modelling as a problem-solving activity in a PBL setting, it can be a powerful learning experience towards actualising their mathematical reasoning abilities while making meanings of specific situations. As both mathematical modelling and PBL are relatively new approaches in the local (Singapore) primary mathematics classrooms, it is essential to elaborate on them before discussing task design and pupils' solutions.

2 Mathematical Modelling as a Problem-Solving Activity

For all that pupils learn in a mathematics lesson, there should be opportunities for them to apply their curricular and personal knowledge in realistic situations or at least situations that simulate reality. In proposing future-oriented perspectives on problem solving, Lesh and Zawojewski (2007) offered a definition of problem solving that embraces productive thinking: A task, or goal-directed activity, becomes a problem (or problematic) when the "problem-solver" (which may be a collaborative group of specialists) needs to develop a more productive way of thinking about the given situation (p. 782). Thinking in a productive way thus requires the problem solver to interpret a situation mathematically, which usually involves iterative cycles of expressing, testing and revising mathematical interpretations of mathematical concepts drawn from various sources, and Lesh and Zawojewski (2007) contend that mathematical modelling as problem solving is one such approach that treats problem solving as integral to the development of an understanding of any given mathematical concept or process.

Research in mathematical modelling for children stems mainly from the models-and-modelling perspective (Lesh & Doerr, 2003). This perspective asserts that pupils' mathematical knowledge is organized around abstractions and experiences. Knowledge development is seen as an action of the individual with others within a context and with the environment. Pupils develop models as conceptual systems consisting of elements of ideas, concepts, constructs, the relationships between such ideas, and the operationalization of the elements. These models are made explicit through representational media when pupils paraphrase, explain, draw diagrams, categorize, find relationships, dimensionalize, quantify, or make predictions through mathematizing (Lesh, Yoon & Zawojewski, 2007). In other words, when pupils engage in modelling activities, their "(internal) conceptual systems are continually being projected into the (external) world" (Lesh & Doerr, 2003, p. 11) thus making visible their sense-making systems of mathematical reasoning in the form of a variety of representational media such as spoken language, written symbols, graphs, diagrams, and experience-based metaphors. In this sense, what the pupils develop are complex artifacts or conceptual tools needed for some purposes or to accomplish some goal (Lesh, Yoon & Zawojewski, 2007). Recent research has shown that young children can manage situations that involve modelling and have produced models that revealed informal understandings of variations, aggregation and ranking of scores, inverse proportion and weighting of variables (English, 2006; English & Watters, 2005).

3 PBL as an Instructional Approach

PBL has been advocated as a reformed-oriented mathematics classroom practice by some researchers (Chan, 2008; Hiebert et al., 1996; Erickson, 1999). In Singapore, the Ministry of Education has also featured PBL as a pedagogy that supports engaged learning (Ministry of Education, 2005).

The PBL setting is deemed as an appropriate platform to carry out mathematical modelling activities (Hjalmarson & Diefes-Dux, 2007). Since the context of this chapter is written with children as the problem-solvers, the PBL instructional approach is designed for short-cycle sessions with limited data mining unlike PBL in higher institutions of learning where it is the curriculum itself. PBL as an instructional approach has three important tenets, namely, the modeling task, the teacher and the pupils (see Figure 1).

The interaction of these three tenets makes problem-solving meaningful and the learning powerful. The modeling task is problem-driven; providing the stimulation to challenge pupils to generate multiple perspectives from it. The modeling task, a complex problem, requires pupils to work collaboratively to make sense of the task, develop, test and revise their solutions. The teacher functions as a cognitive coach. The teacher can make himself or herself part of the learning community but refrains from providing solutions. This enables the teacher to understand what the pupils know and are able to do. The teacher elicits responses, provides supporting feedback, and can also extend the pupils' thinking by challenging them further about their solutions. In this light, PBL helps develop pupils' self-directedness in learning as the teacher does not offer prescriptive help or can be absent from the group most of the time. According to Tan (2007), these key features of the PBL approach provide a learning environment where cognitive immersion takes place.

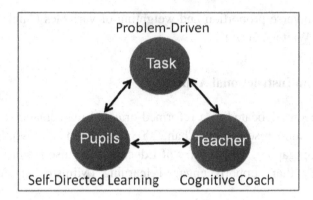

Figure 1. Core tenets of a PBL instructional setting

4 Task Design

Problem-based tasks are characterized as having real-world and curriculum relevance, multiplicity of views, integration of disciplines, ill-structuredness of context, challenge and novelty (Tan, 2003). The features of modelling tasks are similar in that they are both situated in realistic, meaningful contexts that require integration of knowledge and skills toward the solution of the task (Hjalmarson & Diefes-Dux, 2007). Some design principles from selected literature for designing modelling tasks are presented.

The design guidelines from Barbosa (2003) are as follows:
1. The activity must have reference in reality. Although it can be a simulated activity, it is designed to lead to rich discussions and math learning.
2. The activity provides opportunities for pupils to create and pose questions, search for information, organize, manipulate, test, and revise the information.
3. The activity invites pupils to use mathematical ideas, concepts and algorithms and in the process acquire new knowledge.

The following instructional design principles from the works of Lesh et al. (2003) can be used as criteria as well in the designing of mathematical modelling tasks:

1. ***The reality principle***—does the situation resemble some real life experiences and warrant sense-making and extension of prior knowledge?
2. ***The model construction principle***—does the situation create the need to test, modify or extend a mathematically significant construct?
3. ***The self-evaluation principle***—does the situation require pupils to self-assess?
4. ***The construct documentation principle***—does the situation require pupils to make visible their thinking about the situation?
5. ***The simple prototype principle***—is the model adequately simple to use?
6. ***The model generalization principle***—does the solution serve as a useful model for interpreting other similar situations?

The design guidelines and principles are useful to help determine what counts as a modelling activity in a PBL setting. When these characteristics are taken into consideration, the modelling tasks can enable pupils to reveal their knowledge of relevant mathematical ideas as pupils make meanings of real-life situations.

5 Designing Modelling Tasks in a PBL Setting—Two Samples

Two samples of mathematical modelling tasks based on the modelling guidelines and principles described earlier are presented here. For the purpose of illustrating some modelling outcomes, two sample solutions from pupils for each of the modelling tasks are presented. These pupil representations of the modelling tasks represent final solutions and do not include the interactions between pupils or the scaffolding provided by the teacher during the activity.

5.1 *Modelling task 1: The Height-Volume Problem*

A teacher posed a problem to the class. He asked how the height of the water level is related to the volume of the water in a container when water is being poured into different types of containers as shown. "Your friend John believes that the height of the water level is directly related to the volume of water poured in. Your task is to explain clearly that your friend's argument is not good enough. You are to come up with a better explanation to show the relationship to convince the class and the teacher".

The height-volume problem shown in Figure 2 adequately meets the design principles criteria presented earlier.

1. **Real-world setting** (*reality*) — the context takes the form of a case where the pupils need to convince their classmates if the argument put forth in the case is true. The filling of the different containers with water is a scenario pupils are familiar with. Pupils do come across or have experiences with filling of vases, mugs, pails, fishbowls or tanks with water before.

2. **Question posing** (*model construction*) — the pupils will have to ask themselves if the water level and the volume have any relationship. If so, do they behave in the same way for the different shaped containers or do they behave differently?

3. **Making assumptions** (model construction & self evaluation) — to prove the situation, the pupils need to make some assumptions first. For example, they can hypothesize how the water level should behave if the same amount of water is poured into different containers.

4. **Establishing relationships** (model construction) — the pupils will need to generate their own data to show the relationship between the water level and the volume of the water in each different container.

5. **Presenting the model** (construct documentation) — the pupils can present the behaviours of the variables in the form of a table or as a graph.

6. **Interpreting the model** (simple prototype) — the pupils can describe the relationship of the water level and the volume pertaining to the different shaped containers.

7. **Transferring the idea** (generalization) — if any other containers of different shapes are used, the pupils can similarly depict and interpret the relationship between the variables via a graph or the use of hypothetical figures.

Mathematical Modelling in a PBL Setting for Pupils: Features and Task Design   119

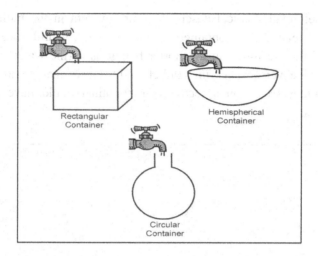

Figure 2. Height-Volume Problem

In trying to make sense of and argue the case, various mathematics concepts are required such as the measurement ideas of height and volume. They have to visualize how the height changes when volume changes in the different containers. They have to use hypothetical data to chart the changes and present the relationships. They need to explain the behaviours of the change and thus justify why the argument printed in the case is not good enough.

The following are two samples of Primary 6 pupils' solutions (Figures 3 and 4 respectively) that reveal the pupils' mathematical conceptual representations of their modelling endeavours.

<u>Solution Sample 1</u>

As seen in Figure 3, the modelling task brought out quite comprehensively the way the pupils conceptualized the situation. The pupils stated that a relationship existed positively, "As the volume increases, the height also increase". They expanded on that relationship by arguing that "BUT the height does not increase at the same rate in all the containers because the widths of the containers are different!" To support their case, they drew two different cuboids and assumed the same volume of water was poured in. The cases yielded different water levels. Their mathematical descriptions are evident in the bullet points where they related that containers that were "wide would have a larger base area" thus resulting in a "shorter height as it (the water) is spread out" and vice versa. The final bullet point shows mathematically the same volumes based on a rectangular container could have different water levels.

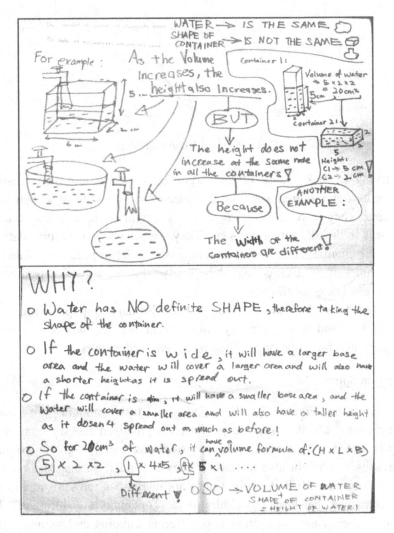

Figure 3. Pupils' solution 1 for the Height-Volume Problem

This group of pupils conceptualized three different relationships of height and volume. The relationships are shown graphically in Figure 4. Their exemplification with respect to the rectangular tank as "...the volume of water increase proportionately to the height" is accurate and they represented the relationship appropriately as linear graph (1). Their descriptions of the height-volume relationships with respect to the second and third containers were not accurately expressed (they could not find the right words) but in the form of graphs, the behaviours of the relationship between height and volume were plausibly depicted for the hemispherical container (2) and the spherical container (3).

<u>Solution Sample 2</u>

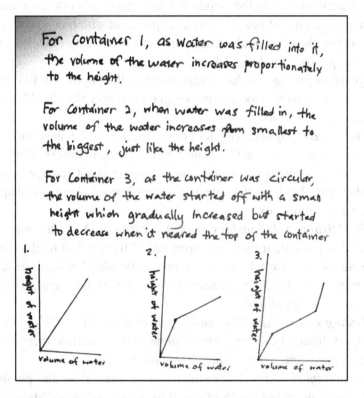

Figure 4. Pupils' solution 2 for the Height-Volume Problem

5.2 Modelling task 2: The Biggest Box Problem

> In a box-making competition, you and your team members are given a sheet of hard paper, rulers, scissors and tapes. Your team is to make the biggest box that can fill the most amount of popcorn.

Figure 5. The Biggest Box Problem

The Biggest Box Problem shown in Figure 5 is considered a modelling task as it adequately meets the design principle/guide criteria, as follows.
1. **Real-world setting** (*reality*) — the context takes the form of a competition where the pupils actually have to work as a team to create the biggest box. Pupils can experiment with how much sand or popcorn their box could contain in comparison with other boxes that other pupils have designed.
2. **Question posing** (model construction) — the pupils will have to ask themselves how a box can be designed and how the variables would affect the volume.
3. **Making assumptions** (model construction & self evaluation) — the pupils may have to define what a box is in their context. They might want to suggest why a cylinder or a pyramid is not considered a box. They may have to give specific interpretations about the shapes of the base of the box.
4. **Establishing relationships** (model construction) — the pupils will need to make their own measurements of the sheet of hard paper and generate their own data to show the relationship between the length, breadth and height to obtain the volume of the box. They may construct a net of a box.
5. **Testing and revising the model** (self evaluation) — the pupils have to test different combinations of dimensions to narrow their range of the dimensions to attain the biggest volume.
6. **Presenting the model** (construct documentation) — the pupils can describe the relationship of the dimensions using a systematic list

and highlight the one that gives the biggest volume or using a graph. Their physical model will be the box they have constructed.

7. ***Transferring the idea*** (generalization) — To find the biggest volume of a box given any dimensions of a piece of square or rectangular board, they may want to consider that the systematic tabulation or graphing method be used to mathematically achieve their goal instead of cutting out box after box.

In the box-making modelling task, the pupils will have to deliberate on mathematical ideas related to geometry and measurement. They should plan a suitable net for the making the box and also realize that only a square box would be possible. They should engage in conjecturing about whether a thin and tall box or a broad and shallow box would give the bigger volume. They will also relate how the lengths of the square cut-outs would affect the volume of the box and work towards plausible dimension combinations to getting the biggest box. As an extension, the pupils should also discuss if examining measurements in decimal fractions would lead them to even bigger volumes.

Solution Sample 1

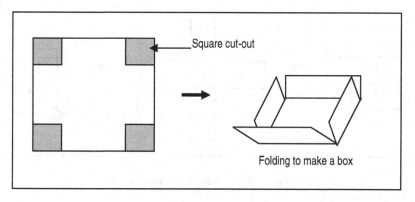

Figure 6. The making of a box (I)

The pupils conceptualized how to make an open box by cutting away four square corners and then folding the four faces as shown in Figure 6. In this sense, they conceptualized the net of a box. As no dimensions were given in the task sheet, the pupils made measurements of the dimensions of the coloured sheet of hard paper. However, to determine how to get the biggest box, they explored the dimension combinations of the length, breadth and height by computing the volume for a cuboid. They adopted a systematic listing of the dimension combinations and realized that as the length of the cut-out increased, the volume of the box was found to increase but up to a certain point before it decreased. The listing of dimension combinations became a modelling tool for the pupils to test the dimensions and volumes. This led to enhancing the volume of the box as they narrowed the dimension range further through considering the decimal fractions of the dimensions. This consideration was assisted through the teacher's promptings and the pupils were able to increase the volume of the box marginally. In this example, the pupils did not use a graphing method but through their mathematical reasoning of tracking the dimension-combination listing, they had exemplified their workings towards obtaining an "ideal" dimension combination that they claimed give them the biggest volume.

Solution Sample 2

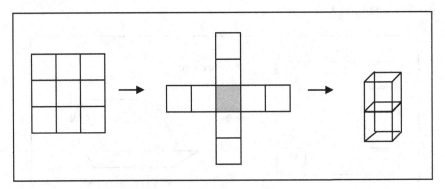

Figure 7. The making of a box (II)

In the sample solution shown in Figure 7, the pupils wanted to make a box such that no materials (square cut-outs) would be wasted. Relying on their understanding of nets, they devised a creative way of getting a volume they thought would be the biggest. They divided the sheet of hard paper into 9 equal squares. They conceived that they could make a net of a box with 9 squares. They used one square as the base area (shaded square in middle picture of Figure 7), and the other 8 squares as the faces of the box with 2 squares per face. They then used the formula of L x B x H to get the volume. This is quite a creative method that is different from most groups that adopted the method used in sample solution (I). The learning value here is not just in terms of the conceptualizing the representation and making the mathematical translation but in comparing and reasoning why solution (II) and solution (I) yielded volumes that are so close despite solution (I) having involved wastage of materials.

6 Concluding Remarks

In this chapter, mathematical modelling is viewed as a problem-solving activity. This activity facilitates social learning and uses modelling tasks to drive the learning. It is situated in the PBL instructional setting where the interaction between pupils, task and the teacher is posited to provide a learning environment where cognitive immersion takes place. From a modelling perspective, the focus is on the process of developing conceptual representations and mathematical descriptions of specific situations. For pupils to construct models as representations through relating variables and operationalizing them, modelling tasks have to be appropriately designed to meet the goals of enabling pupils to acquire problem solving skills and habits for real-life applications. Modelling tasks used in a PBL setting have to have features that embed modelling and problem-solving aspects. As such one should expect that a search for a solution would not be as structured and tidy like in solving textbook word problems where there are assured ways of finding the solution which involve neat numerical figures. Teachers can design meaningful modelling tasks for pupils to investigate and solve by gaining an understanding of what counts as a modelling task and embracing some modelling principles as guides in the design of the task. The modelling tasks could be ill-structured and have rich data with the context embedded in reality. The tasks should enable pupils to express what they have in mind, test their conjectures and revise them accordingly. In other words, they must enable the pupils to continually question their representations or designs towards refining them in the direction of attaining a workable solution model for a specific purpose. In so doing, pupils should learn to transfer their modelling knowledge to other similar situations.

It must be noted however that designing good modelling activities is only one side of the coin. Knowing how to execute the modelling activities has to be considered as well for learning to be maximized. To reap the benefits of mathematical modelling and PBL, such activities should become a common problem-solving feature in the mathematics classroom. Carrying out such activities once or twice would not be helpful to acclimatize both pupils and teachers in the spirit of learning via problem solving meaningfully. Pupils and teachers need time to get acquainted with team members' dispositions and styles and also with the type of tasks at hand since they are different from the conventional classroom practices. The teachers should also learn to acquire some facilitation skills that would benefit the pupils' modelling process.

As notions of problem-solving research are advancing alongside changing educational landscape, the emphasis on what is needed from pupils in problem solving has changed to keep in line with new skills and competencies for the 21st century. Situating mathematical modelling in a PBL setting holds promise as an excellent platform towards developing those abilities of mathematical thinking that involve creating and interpreting situations. Designing modelling tasks based on the design principles or guides is a step towards enabling pupils to engage in meaningful problem solving.

References

Barbosa, J. C. (2003). What is mathematical modelling? In S. J. Lamon, W. A. Parder &. K. Houston (Eds.), *Mathematical modelling: a way of life* (pp. 227-234). Chichester: Ellis Horwood.

Chan, C. M. E. (2008). Using model-eliciting activities for primary mathematics classrooms. *The Mathematics Educator, 11*(1/2), 47-66.

English, L. D. (2002). Priority themes and issues in international mathematics education research. In L. D. English (Ed.), *Handbook of international research in mathematics education* (pp. 3-16). Mahwah, New Jersey: Lawrence Erlbaum Associates.

English, L. D. (2006). Mathematical modelling in the primary school: Children's construction of a consumer guide. *Educational Studies in Mathematics, 63*, 303-323.

English, L. D., & Watters, J. J. (2005). Mathematical modelling in third-grade classrooms. *Mathematics Education Research Journal, 16*(3), 59-80.

Erickson, D. K. (1999). A problem-based approach to mathematics instruction. *The Mathematics Teacher, 92*(6), 516-521.

Hiebert, J., Carpenter, T. P., Fennema, E., Fuson, K., Human, P., Murray, H., Olivier, A., & Wearne, D. (1996). Problem solving as a basis for reform in curriculum and instruction: The case of mathematics. *Educational Researcher, 25* (4), 12-21.

Hjalmarson. M. A., & Diefes-Dux, H. (2007). Teacher as designer: A framework for teacher analysis of mathematical model-eliciting activities. *The Interdisciplinary Journal of Problem-based learning, 2*(1), 57-78.

Lesh, R., & Doerr, H. M. (2003). Foundations of a models and modeling perspective in mathematics teaching and learning. In R.A. Lesh & H. M. Doerr (Eds.), *Beyond constructivism: Models and modeling perspectives on mathematics problem solving, learning and teaching* (pp. 3-34). Mahwah, NJ: Lawrence Erlbaum Associates.

Lesh, R., & Lehrer, R. (2003). Models and modelling perspectives on the development of students and teachers, *Mathematical Thinking and Learning* 5(2&3), 109–130.

Lesh, R., & Zawojewski, J. (2007). Problem solving and modeling. In F. K. Lester (Ed.), *Second handbook of research on mathematics teaching and learning: A project of the National Council of Teachers of Mathematics* (pp. 763-803). Charlotte, NC: Image Age Publishing.

Lesh, R., Yoon, C., & Zawojewski, J. (2007). John Dewey revisited: Making mathematics practical versus making practice mathematical. In R. A. Lesh, E. Hamilton, & J. J. Kaput (Eds.), *Foundations for the future in mathematics education* (pp. 315-348). Mahwah, NJ: Lawrence Erlbaum Associates.

Lesh, R., Cramer, K., Doerr, H., Post, T., & Zawojewski, J. S. (2003). Model development sequences. In R.A. Lesh & H. M. Doerr (Eds.), *Beyond constructivism:*

Models and modeling perspectives on mathematics problem solving, learning and teaching (pp. 35-58). Mahwah, NJ: Lawrence Erlbaum Associates.

Ministry of Education (2005). *Principles of engaged learning: PETALS*. Singapore: Curriculum Planning and Development Division

Tan, O. S. (2003). *Problem-based learning innovation: Using problems to power learning in the 21st century*. Singapore: Thomson Learning.

Tan, O. S. (2007). Using problems for e-learning environments. In O.S. Tan (Ed.), *Problem-based learning in elearning breakthroughs (pp. 1-14)*. Singapore: Thomson Learning.

Verschaffel, L., De Corte, E., & Borghart, I. (1997). Pre-service teachers' conceptions and beliefs about the role of real-world knowledge in mathematical modeling of school word problems. *Learning and Instruction, 7*(4), 339-359.

Chapter 8

Initial Experiences of Primary School Teachers with Mathematical Modelling

NG Kit Ee Dawn

This chapter reports on some initial experiences of primary school teachers when engaged in the processes of mathematical modelling. The teachers worked in groups of four on one out of three model-eliciting tasks targeted at upper primary levels and then presented their approaches. The approaches used for each task were based on the assumptions and conditions imposed by the groups. Peer evaluation of the mathematical models was conducted based on three criteria: representation, validity and applicability. Although there was enthusiastic participation in these tasks, it seems that more scaffolding is crucial to ease teachers into the implementation of such tasks in the primary classrooms, particularly in nurturing the mindset change of teachers towards accepting multiple representations of a problem, diverse solutions, and what constitutes as mathematical in the representations.

1 Introduction

Mathematical modelling was incorporated into the revised primary school mathematics syllabus in Singapore in 2003. In the revised syllabus, mathematical modelling refers to the "process of formulating and improving a mathematical model to represent and solve real-world problems" (Ministry of Education, 2006, p. 14). In the syllabus, it is also stated that the objective of mathematical modelling is to allow students to "use a variety of representations of data", and to "select and apply appropriate mathematical methods and tools" (p. 14). Three key features can be identified in a modelling process: (a) representation, (b) validity, and (c) applicability (Kaiser, 2005). During the modelling process, a student interprets a given real-world problem within his or her socio-cultural context and forms his or her own mental image of the problem at hand. Next, various possible perspectives at representing the problem mathematically focusing on factors affecting the outcome of the problem and the relationships between the selected factors and the outcome of the problem are considered. A mathematical model is generated (representation). The solution generated by the model is then validated with the real-world situation until the model is perceived to be reasonable at providing a framework at which the problem could be addressed (validity). Very often, this is a cyclical process. Modellers also work towards whether their models can be applied to other related problem situations (applicability).

Many perspectives on mathematical modelling abound. The tasks used for the purposes reported in this chapter adhere to the *"contextual modelling perspective"* (Kaiser & Sriraman, 2006, p. 306) where decision making during the modelling process factors in real-world experiences and integration of socially constructed knowledge across disciplines. This perspective of modelling is chosen because it is of direct relevance to the current focuses of the Singapore mathematics curriculum, in particular the activation of the processes of *reasoning, mathematical communication*, and *connections* during problem solving. Specifically, connections refer to linkages within mathematical topics, between mathematics and other subject-disciplines, and between mathematics and real-life experiences.

Proponents of this perspective (e.g., Chan, 2008; English, 2009; Iversen & Larson, 2006; Lesh & Doerr, 2003; Lesh & Sriraman, 2005a, 2005b) espoused the use of *model-eliciting activities* which involve students "making sense of meaningful situations", as well as attempting to "invent", "extend", and "refine their own mathematical constructs" (Kaiser & Sriraman, 2006, p. 306). Students first formulate models for solving a problem embedded in a real-world context. The model is then applied to a new related problem. In doing so, students are motivated to naturally develop the mathematics needed for representing given situations. However, for the context of work reported in this chapter, the contextual modelling perspective cannot be taken in its purest sense. This is because the experiences of 48 primary school *teachers* in model-eliciting tasks targeted at grades 5 and 6 will be analysed. Instead of the need to develop the mathematics brought about by the tasks, it is assumed that the teachers already have the necessary mathematical knowledge and skills the tasks required to complete the tasks. Of interest to the author in this preliminary investigation is the nature of mathematical models elicited, the challenges faced by the teachers in their initial exposure to such activities, and most of all, the conceptions of teachers about such tasks being a part of the teaching and learning cycle.

2 Potentials and Challenges to the Use of Real-World Contextualised Tasks

As a type of "reality-based contextual examples", modelling problems are espoused to play an important role in mathematics education (Kaiser, 2005, p. 2). Studies have shown that mathematical tasks embedded within meaningful real-world experiences of students engaged students in learning (Blum, 2002; Sembiring, Hadi, & Dolk, 2008; Van den Heuval-Panhuizen, 1997, 1999). Shaped by students' own relations with them (Lave & Wenger, 1994), such tasks create opportunities for flexible, adaptive applications of taught or self-discovered knowledge and skills determined by students' personalised interpretations of the context.

Nonetheless, challenges to producing *quality mathematical outcomes* from contextualised tasks exist. Some of these challenges could be due to differing reactions to the contexts brought forth by these tasks. Two broad interpretations to contextualised tasks are proposed by Clarke and Helme (1996): (a) context as referred to but bounded by task descriptions and (b) real-life context in which the task is crafted out of. The second interpretation involves a certain amount of extrapolation of details inferred from the tasks. Contrary to intentions one of the challenges faced by proponents of contextualised task is that students can choose not to engage with the mathematical aspects afforded by the task, creating realistic but non-mathematical solutions (Gravemeijer, 1994). Such observations were echoed by Busse (2005) who discovered that some students in his sample did not mathematise the given problem nor apply mathematical methods when dealing with contextualised problems. He postulated that these students considered a contextualised task given in classroom situation to be fully descriptive of an actual real-world situation (i.e., the second interpretation above) which in itself called for applications of socio-cultural norms for decision making over and above school-based learning. In other words, students may not see the need to represent the given problem from the mathematical perspective when they can interpret it using their integrated view of the world. Busse's (2005) findings matched those from Venville, Rennie, and Wallace (2004) as well as Ng (2008). The former found that high school students did not consider knowledge from subject-specific disciplines as useful as expected nor have they used their prior learning as often as desired during contextualised tasks. In the latter study, several members from the student-groups in grades 7 and 8 chose not to activate relevant prerequisites from their mathematics lessons during a design-based task involving concepts of mensuration and scale measurement, opting to approach the task in non-mathematical ways.

On the other hand, researchers (e.g., Ng & Stillman, 2009; Verschaffel, Greer, & De Corte, 2000) found that some students ignored the contexts provided and did not activate their "real-world knowledge and realistic considerations" (Van den Heuval-Panhuizen, 1999, p. 137) for decision making, verification, and sense making during mathematical knowledge application across a variety of task and cultural situations. Verschaffel et al. (2000) referred to this as the "suspension of meaning" (p. 3) following Schoenfeld's (1991) critique of students' failure to use real-world knowledge and their sense making facilities in their problem solving attempts. Busse (2005) used the concept of "mathematics bound" (p. 355) to describe such cases. Students belonging to this category are said to view contexts as superficial enhancements of the mathematics to be applied in the problem. These students are focused mainly on mathematical concepts and skills and how to apply the mathematics they know to the problem. Thus, they are more interested in the direct and immediate translation of information provided into mathematical expressions before proceeding to solving the problem with little consideration of real-world constraints as insinuated by the context.

In addition, there are two other challenges during the implementation of contextualised tasks. The first of which is that against assumptions students may not find the context provided in a given task meaningful to them for a large variety of reasons. For one, some students may have different life experiences from others and hence are unable to relate to the context given (Stillman, 2000), much less interpret the task in a realistic way. Others may find the task context unappealing (Norton, 2006). A second challenge comes from students' lack of ability to monitor the quality of their own mathematics applications and those of their peers in the groups during participation in contextualised tasks (Ng, 2008; Wong, 2001). Even when subject-based knowledge is used, students could still face difficulty in monitoring the appropriateness and accuracy of application of concepts and skills. This has been and will continue to be the concern of teachers implementing such tasks.

3 Significance of Study

Although much has been investigated on students' reactions to contextualised tasks, there was limited documentation on how teachers themselves view such tasks. It is of interest whether the teachers face the same challenges in these tasks as reported above during their initial exposure, and how they approach the tasks given their professional sensitivity to the mathematical affordances of the tasks, particularly when given tasks are attuned to the conceptual levels of their students. One study by Verschaffel, De Corte, and Borghart (1997) investigated the beliefs and conceptions of 332 pre-service primary school teachers about the role of real-world knowledge in a problem context through the teachers' spontaneous responses to a set of word problems with problematic modelling assumptions. The researchers concluded that teachers' cognition and beliefs about the role of real-world knowledge in the interpretation of problems have strong implications on their actual teaching behaviours and consequently on students' learning. Two key findings attributed to this conclusion. Firstly, almost half of their sample excluded the use of real-world knowledge in their solutions to problems within primary to secondary schooling range of mathematical content. When asked to comment on responses to the same tasks from actual students, some of these teachers were unable to critically examine the fallacy in realistic mathematical meaning-making within the contexts provided. Secondly, it was a common belief among these pre-service teachers that grade five students should not be confronted with such complex, ill-structured, or tricky problems because the goal of teaching word problem solving was to elicit the necessary mathematical concepts and skills from the problem and work towards producing a numerical response.

In Singapore, there is little research into teachers' use of mathematical modeling tasks. Documenting the initial experiences of teachers to these tasks paves the way for enhanced teacher training towards the facilitation of contextualised tasks. Mathematical modelling is a new activity only introduced in the latest syllabus (Ministry of Education, 2006). It is understandable that teachers would be curious but perhaps apprehensive about what it involves for them as facilitators and for the students as learners in modelling tasks. As Wong (2008) proposed, "context knowledge" (p. 2) should be included as another component in the Singapore mathematics curriculum. Teachers should be aware of how context plays a part in teaching and learning of mathematics, specifically, the careful choice of contexts may enhance or impede students' appreciation of mathematics. Furthermore, in relation to representation, validity, and applicability of the modelling process, it would be interesting to study the approaches used by teachers in similar kinds of modelling tasks for their students, given their vastly superior mathematical knowledge and pedagogical experience. These findings will have implications for teacher training on the facilitation of modelling tasks for better quality mathematical outcomes.

4 Method

The preliminary study reported here was carried out during a 3-hour interaction period between the author and 48 primary school teachers in Singapore. Each participant had at least one year of teaching experience in schools.

4.1 *Data collection and analysis procedure*

The teachers first attended a 20-minute introductory lecture on the definition of mathematical modelling and its key features. Next, they were asked to work in groups of four to formulate a mathematical model for an assigned model-eliciting task. They were given an hour to do it. There were altogether three tasks prepared for the session, with four groups working on each task. Every group recorded their mathematical model on a flip chart along with assumptions, conditions, and variables considered. A gallery walk consisting of poster presentations of written work done was conducted at the end of the group discussions. Every group either appointed one presenter to present their work thrice or assigned members to take turns to present their work. The presentations were conducted as three repeat concurrent sessions where group members who were not presenting were free to move around from one poster presentation to another. Each teacher was encouraged to attend at least two presentations on a task other than their own during the gallery walk.

As the teachers listened to the presentations, they wrote down their comments based on three criteria: representation, validity, and applicability using a reflection sheet. The reflection sheet consisted of three questions for consideration about the model formulated:

(a) representation - how well does this model solve the problem?

(b) validity - can you suggest how to improve upon the model?, and

(c) applicability - can this model be used in other contexts?

As this was the first time the teachers in the sample encountered mathematical modelling and given the short interaction time, they were not be expected to conduct in-depth discussions nor produce evidence about validity and applicability testing attempts for their models.

Throughout the group discussions, the author went from group to group, taking note of the queries and concerns of the teachers as they embarked on the tasks. After the gallery walk, the author summarised the approaches used for each task, addressed common concerns and difficulties of the teachers about the tasks, and discussed the models presented in terms of the three criteria. At the same time, she invited the teachers to comment on the use of such tasks in their own classrooms. All written work done and reflection sheets were collected at the end of the session. The mathematical approaches used for each task were analysed along with comments on the representation, validity, and applicability of the mathematical models by the author. In addition, field notes on teacher difficulties with the tasks and their conceptions of such tasks were examined. One of the tasks and findings related to it will be elaborated below. This task was attempted by Groups 5, 6, 7, and 8.

4.2 Task: Youth Olympic Games

Singapore will host its first Youth Olympic Games (YOG) in August 2010. Preparations for YOG were underway when the teachers participated in the task (Appendix). Adapted from English (2007) who used hers with grade 5 students, the YOG task required the groups to propose a set of selection criteria for two female swimmers in the Women's 100m freestyle event during YOG. A table displaying the results of five candidates in 10 most recent 100m freestyle competitions was given to facilitate their mathematical approaches to the problem. This problem was chosen because there were multiple ways of interpreting it (e.g., using average, by ranking) as well as vast opportunities for the groups to make decisions on the variables (e.g., best time for each competitor, consistency in performance) to focus upon for the selection criteria, subsequently determining the types of mathematics which can be applied.

5 Findings

5.1 *Mathematical approaches used*

Three out of four groups (Groups 5, 6, and 8) who participated in this task modelled the problem using *table representations* of values. All three groups chose *average timing* as a common criterion for selection of swimmers for YOG, with Groups 6 and 8 treating it as their only criterion. Nevertheless, the two groups differed in their approaches. Instead of using the timings for all 10 competitions for the calculation of average timing for each competitor, Group 6 chose to work with the average timing of the three most recent competitions. In contrast, as seen in Table 1, Group 5 attempted to provide a more holistic set of criteria for selection. The group even recommended an additional criterion on Sportsmanship Attitude where each competitor was to undergo an interview to be graded on the degree of professionalism in sports. However, as in Groups 7 and 8, the relevant mathematical calculations were not complete in the model presented by Group 5. It appeared that the teachers from at least three of the groups had only managed to outline their approach rather than to really work through the task mathematically. This could explain why Group 5 was not able to explain how they would make use of their five criteria in an integrated way to select the swimmers. In addition, it was not clearly indicated in the work done how average timing for each competitor was calculated as each competitor did not participate in all 10 previous competitions.

Group 7 was the remaining group among the four who tried using *graphical representation* for the data provided, albeit incomplete (Figure 1). They meant to plot a line graph showing the time taken for each swimmer against the 10 competitions recorded. They had wanted to compare the trend of time taken from the first to the tenth competition for individual swimmers, hoping to base their selection on the best two swimmers who showed consistency in improvement. However, it was uncertain whether they had discussed how they would make their selections if almost all the swimmers had both upward and downward trends in their timings throughout the 10 competitions.

Table 1

Table representation by Group 5

Criteria	A	B	C	D	E
Improvement in timing from first to last competition (seconds)	3	8	-5		
Average timing from previous competitions (seconds)	58.13	56.00			
Total number of previous competitions missed	2	3	4		
Time in most recent competition (minutes and seconds)	00:56	00:49	01:02		
Scores on Sportsmanship Attitude					

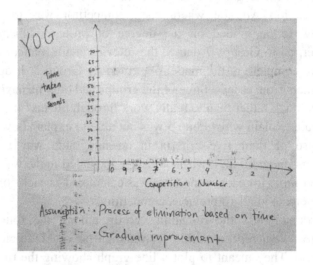

Figure 1. Graphical representation by Group 7

5.2 Comments from peers

Peer evaluation of work from the four groups who participated in the YOG task was mainly positive in terms of how well the formulated models answered the question (representation). Overall, in terms of applicability of the models, the teachers suggested other contexts the models could be used for (e.g., sports day in the school, 100 m sprint events). In particular, the teachers agreed with Group 6 that the average timing for the three most recent competitions for each swimmer should be used as a criterion for selection. However, many cautioned that in terms of validity of the model in addressing the problem, the group had to consider other factors like (a) not all swimmers competed in all prior events, (b) stamina of swimmers in competing race after race, (c) health conditions of the swimmers, and (d) reasons for certain swimmers having not competed in certain competitions.

5.3 *Real-world knowledge activated and reasonableness checked*

It would seem that the nature of the task, the use of poster presentations, and the structure of the reflection worksheet facilitated the activation of real-world knowledge among the teachers for choice of model representation and checking of accuracy and reasonableness of approach. All the groups enthusiastically discussed the information presented as well as those missing from the tabulated results provided in the task. They could see the task as being meaningful and the potential of it being adapted for their use in the classroom. The teachers not only analysed the information, classified them, but also formed criteria for selection of swimmer such as number of competitions participated in and how recent were the results of prior competitions. In addition, the teachers critically questioned the nature and sufficiency of information provided in the task. For example, teachers from Group 5 explained they needed more information on the attitudes of the swimmers because in real-life, competitors can be banned from participating in events because they lacked integrity and professionalism even though they may be the top few in their field. By requesting other teachers to comment on the work done by the four groups, more discussions were generated on the reasonableness of choice of criteria and other possible criteria not considered which are pertinent in the real-world context. This could be seen from their suggestions to improve the validity of the model presented by Group 6 which is explained above.

5.4 *Challenges faced by the teachers*

Nonetheless, it could be observed that the teachers faced several challenges in their attempts at the task. For example, with the exception of Group 6 who recommended swimmers B and D due to their comparatively faster average times for the three most recent competitions, the other groups did not manage to select their swimmers based on their models. One of the reasons cited was that the calculations associated with their choice of approach were too tedious. Given their understanding of the problem, they were satisfied with brief descriptions of their approaches in writing since they could elaborate on them verbally during the poster presentations.

In addition, some groups needed guidance in getting started on the task. To them, the task did not allow for direct application of algorithms in a structured manner they were comfortable with. There were too many variables to consider and there was no fixed approach or answer to the problem. For example, members in Group 8 spent a lot of time deciding on possible criteria to use for the selection of swimmers based on the tabulated values presented in the task. They subsequently decided to simplify the problem by just considering the average timing for each swimmer. Even then, they were apprehensive after the decision was made because they were not confident whether they were on the 'right track'.

Moreover, collectively the teachers in each group displayed very brief written communication about their mathematical reasoning while attempting the task. Some indicated that they preferred the oral mode of delivering detailed explanations of their approaches. Others shared that they had difficulty communicating their reasoning processes in lengthy written form as they were not used to this. In addition, the need to state the assumptions, conditions, and variables considered for the task posed challenges to the teachers. It was not easy to differentiate between the terms, much more to explain them clearly to their students.

Furthermore, from the reflection sheets, it appeared that the teachers not only needed time to adjust to critiquing in writing their peers' work but also to do so in formats other than what they were used to, although questions was provided to scaffold considerations of aspects of representation, validity, and applicability of the formulated models. They mainly wrote brief comments on the presentation of work they observed and seemed to have almost totally avoided discussions of mathematical concepts and skills (see Figures 2a and 2b).

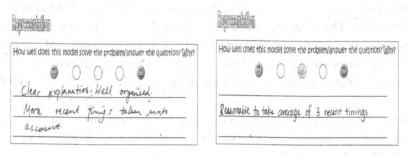

Figure 2a. About the work of Group 8   Figure 2b. About the work of Group 6

5.5 Beliefs and conceptions about the task and modelling activities

Three observations about the beliefs and conceptions of teachers could be obtained from the teachers' work in the YOG task. Firstly, the teachers believed in extending their interpretations of the task to the real-world context it represented. Subsequently, relevant real-world knowledge was activated in mathematical decision making (e.g., Group 5) and in the critique of the work of others. Secondly, some teachers (as observed in Group 8) seemed to have maintained a fixed mindset that mathematical representations should only be in the form of structured algorithms and formulae during problem solving. As a result, they had initial difficulties trying to 'accept alternative representations' such as tabulation of values, graphs, and listing so much so that they wondered whether these representations were 'mathematical enough' to be part of the solution processes. These teachers also expressed concerns about facilitating students on similar tasks, especially the evaluation of work. Last but not least, some teachers believed that there should only be one answer to the task given and were therefore always on the lookout for the 'correct' approach and a unique answer. Hence, there were feelings of apprehension and uncertainty throughout their experiences with the task, having the need to seek reaffirmation that they were on the 'right track'.

6 Discussion and Conclusion

Drawing upon the cautionary advice from Stillman (2000) and Norton (2006) on the impact of task design on meaningful learning, the choice of YOG task appeared to be appropriate because the teachers in the sample had some shared knowledge of the event due to publicity of the Games in Singapore. The resulting enthusiastic participation of the teachers in the task revealed interesting findings which extended from those of previous studies on students' experiences with contextualised activities. For instance, like the students in Gravemeijer's (1994) and Busse's (2005) studies, the teachers also seemed open to the use of real-world knowledge in the task. However, unlike the students from these studies who demathematised their tasks, the teachers here attempted to present what they thought to be mathematical solutions for the task. In addition, as what Busse (2005) and Verschaffel et al. (1997) discovered, there were teachers who were more mathematics bound than others. They sought the immediate translation of information provided in the context of the task into mathematical expressions or known algorithms so as to obtain unique answers to the problem. These teachers showed clear discomfort with the openness of the task.

These findings suggest although the design of model-eliciting activities seem to promote activation of real-world knowledge and provide opportunities for development of reasoning processes and mathematical communication, teacher educators have to bridge the gaps between the potentials of such tasks and teachers' current levels of expertise (i.e., content and pedagogy), as well as raise their comfort levels to such open-ended contextualised tasks. For example, teachers could be invited to use mathematics themselves and practise engaging in open discussions about mathematical concepts and skills verbally or in writing. This will aid their facilitation and evaluation of students' presentation of their reasoning processes and ability to communicate with others mathematically. Moreover, teachers could also be encouraged to share how they used mathematics in real-world contexts, how they select relevant real-world knowledge for mathematical decision making, and perhaps also how they model real-world problems mathematically and check the validity and applicability of their models.

In this way, teachers may inspire their students to do the same and thus encourage a more connected view of school-based learning. Last but not least, slow but deliberate attempts have to be made to promote a mindset change of the teachers in accepting alternative ways of mathematical representations which may result from differing learning activities and preparedness to work outside their comfort zone in facilitating these activities.

Acknowledgement

The author would like to thank Dr Gloria Stillman from the University of Melbourne, Australia for her valuable comments to this chapter.

References

Blum, W. (2002). ICMI Study 14: Applications and modelling in mathematics education. *Educational Studies in Mathematics, 51*(1/2), 149-171.
Busse, A. (2005). Individual ways of dealing with the context of realistic tasks: First steps towards a typology. *ZDM — The International Journal on Mathematics Education, 37*(5), 354-360.
Chan, C. M. E. (2008). Using model-eliciting activities for primary mathematics classrooms. *The Mathematics Educator, 11*(1/2), 47-66.
Clarke, D., & Helme, S. (1996). Context as construction. In C. Keitel, U. Gellert, E. Jablonka & M. Müller (Eds.), *Mathematics (education) and common sense:*

Proceedings of the 47th meeting of the International Commission for the Study and Improvement of Mathematics Teaching in Berlin (pp. 379-389). Berlin: CIEAM.

English, L. (2007). Modeling with complex data in the primary school [Electronic Version]. *Thirteenth International Conference on the Teaching of Mathematical Modelling and Applications.* Retrieved July 22-26, from http://site.educ.indiana.edu/Papers/tabid/5320/Default.aspx

English, L. (2009). Promoting interdisciplinarity through mathematical modelling. *ZDM — The International Journal on Mathematics Education, 41*(1-2), 161-181.

Gravemeijer, K. P. E. (1994). *Developing realistic mathematics education.* Utrecht, The Netherlands: Freudenthal Institute.

Iversen, S. M., & Larson, C. J. (2006). Simple thinking using complex math vs. complex thinking using simple math — A study using model eliciting activities to compare students' abilities in standardized tests to their modelling abilities. *ZDM — The International Journal of Mathematics Education, 38*(3), 281-292.

Kaiser, G. (2005, February). *Mathematical modelling in school — Examples and experiences.* Paper presented at the Paper presented at the Fourth Congress of European Society for Research in Mathematics Education, Saint Felix de Guixols, Spain.

Kaiser, G., & Sriraman, B. (2006). A global survey of international perspectives on modelling in mathematics education. *ZDM — The International Journal of Mathematics Education, 38*(3), 302-310.

Lave, J., & Wenger, E. (1994). *Situated learning: Legitimate peripheral participation.* New York: Cambridge University.

Lesh, R. A., & Doerr, H. M. (2003). Foundations of a models and modeling perspective in mathematics teaching and learning. In R. Lesh & H. M. Doerr (Eds.), *Beyong constructivism: Models and modeling perspectives on mathematics problem solving, learning and teaching* (pp. 3-34). Mahwah, NJ: Lawrence Erlbaum Associates.

Lesh, R., & Sriraman, B. (2005a). John Dewey revisited — Pragmatism and the models-modeling perspective on mathematical learning. In A. Beckmann (Ed.), *Proceedings of the 1st International Symposium on Mathematics and its Connections to the Arts and Science* (pp. 32-51). Franzbecker, Hildesheim: Pedagogical University of Schwaebisch Gmuend.

Lesh, R., & Sriraman, B. (2005b). Mathematics education as a design science. *ZDM, 37*(6), 490-505.

Ministry of Education. (2006). *Mathematics Syllabus (Primary).* Singapore: Author.

Ng, K. E. D. (2008). *Thinking, small group interactions, and interdisciplinary project work.* Unpublished doctoral dissertation, The University of Melbourne, Australia.

Ng, K. E. D., & Stillman, G. A. (2009). Applying mathematical knowledge in a design-based interdisciplinary project. In R. Hunter, B. Bicknell & T. Burgess (Eds.), *Crossing, divides: Proceedings of the 32nd annual conference of the Mathematics Education Research Group of Australasia* (Vol. 2, pp. 411-418). Wellington, New Zealand: Adelaide: MERGA.

Norton, S. (2006). Pedagogies for the engagement of girls in the learning of proportional reasoning through technology practice. *Mathematics Education Research Journal, 18*(3), 69-99.

Sembiring, R. K., Hadi, S., & Dolk, M. (2008). Reforming mathematics learning in Indonesian classrooms through RME *ZDM — The International Journal on Mathematics Education, 40*(6), 927-939.

Schoenfeld, A. H. (1991). On mathematics as sense-making: An informal attack on the unfortunate divorce of formal and informal mathematics. In J. F. Voss, D. N. Perkins & J. W. Segal (Eds.), *Informal reasoning and education* (pp. 311-343). Hillsdale, NJ: Lawrence Erlbaum Associates.

Stillman, G. (2000). Impact of prior knowledge of task context on approaches to applications tasks. *Journal of Mathematical Behavior, 19*(2000), 333-361.

van den Heuvel-Panhuizen, M. (1997, September). *Realistic problem solving and assessment.* Paper presented at the Mathematics Teaching Conference, University of Edinburgh, Scotland.

van den Heuvel-Panhuizen, M. (1999). Context problems and assessment: Ideas from the Netherlands. In I. Thompson (Ed.), *Issues in teaching numeracy in primary schools* (pp. 130-142). Buckingham, UK: Open University Press.

Venville, G., Rennie, L., & Wallace, J. (2004). Decision making and sources of knowledge: How students tackle integrated tasks in science, technology and mathematics. *Research in Science Education, 34*(2), 115-135.

Verschaffel, L., De Corte, E., & Borghart, I. (1997). Pre-service teachers' conceptions and beliefs about the role of real-world knowledge in mathematical modelling of school word problems. *Learning and Instruction, 7*(4), 339-359.

Verschaffel, L., Greer, B., & De Corte, E. (2000). *Making sense of word problems.* Lisse, The Netherlands: Swets & Zeitlinger.

Wong, H. M. (2001). *Project work in primary school: Effects on learning and teaching.* Unpublished master's thesis, Nanyang Technological University, National Institute of Education, Singapore.

Wong, K. Y. (2008). An extended Singapore mathematics curriculum framework. *Maths Buzz, 9*(1), 2-3.

Appendix

Youth Olympic Games Problem

Singapore will be hosting the first Youth Olympic Games (YOG) from 14 to 26 August 2010. It will receive some 3,600 athletes and 800 officials from 205 National Olympic Committees, along with estimated 800 media representatives, 20,000 local and international volunteers, and more than 500,000 spectators. Young athletes - between 14 and 18 years of age - will compete in 26 sports and take part in Culture and Education Programme. The Singapore 2010 Youth Olympic Games will create a lasting sports, culture and education legacy for Singapore and youths from around the world, as well as enhance and elevate the sporting culture locally and regionally. There are altogether 201 events featuring 26 different kinds of sports such as swimming, badminton, cycling, fencing, table tennis, volleyball and weightlifting. Singapore is well-known in the region for grooming young swimmers.

The Singapore Sports School is having difficulty selecting the most suited swimmers for competing in the Women's 100m freestyle event. They have collected data on the top five young female swimmers over the last 10 competitions. To be fair, codes are used to represent the swimmers until the selection process is over. The list of swimmers and their codes are only known to you and your group. As part of the Singapore Sports School YOG committee, you need to use these data to develop a method to select the two most suited women for this event.

Study the data presented below collected over two years (2007-2008). Records for Competition 1 are the most recent. Records for Competition 10 are the oldest.

Women's 100m freestyle Results recorded (seconds)*

Competition No.	Time in 100 m Freestyle (Minutes and Seconds)				
	Swimmer A	Swimmer B	Swimmer C	Swimmer D	Swimmer E
1	00:56	00:49	01:02	00:57	00:55
2	00:55	DNC	00:59	01:05	DNC
3	DNC	00:56	DNC	00:59	00:55
4	00:58	01:01	00:57	DNC	01:04
5	DNC	00:57	DNC	00:58	00:49
6	00:59	DNC	DNC	00:56	00:57
7	01:00	DNC	00:59	00:56	00:57
8	00:59	00:56	00:55	00:58	00:57
9	00:59	00:56	DNC	DNC	00:58
10	00:59	00:57	00:57	DNC	00:58

DNC: Did not compete. *Best time across heats, semi-finals and finals.

(1) Decide on which two female swimmers should be selected.

(2) Write a report to the YOG organizing committee in Singapore to recommend your choices. You need to explain the method you used to select your swimmers. The selectors will then be able to use your method to select the most suitable swimmers for all other swimming events.

[Adapted from English, L. (2007). Modeling with complex data in the primary school [Electronic Version]. *Thirteenth International Conference on the Teaching of Mathematical Modelling and Applications.* Retrieved July 22-26, from http://site.educ.indiana.edu/Papers/tabid/5320/Default.aspx.]

Part III

Applications & Modelling in Secondary School

Chapter 9

Why Study Mathematics? Applications of Mathematics in Our Daily Life

Joseph Boon Wooi YEO

Most students would like to know why they have to study various mathematical concepts. Teachers usually cannot think of a real-life application for most topics or the examples that they have are beyond the level of most students. In this chapter, I will first discuss the purposes of mathematics, the aims of mathematics education and the rationales for a broad-based school curriculum. Then I will provide some examples of applications of mathematics in the workplace that secondary school and junior college students can understand. But students may not be interested in these professions, so I will examine how mathematics can be useful outside the workplace in our daily life. Lastly, I will look at how mathematical processes, such as problem solving, investigation, and analytical and critical thinking, are important in and outside the workplace.

1 Introduction

When teachers try to convince their students that mathematics is useful in many professions, such as engineering and medical sciences, many of their students may not be interested in these occupations. For example, when I was a teacher, some of my students wanted to be computer game designers instead, but they wrongly believed that this profession did not require much mathematics: only when I demonstrated to them that computer programming required some mathematics did they show any interest in studying mathematics. Other students of mine aspired to be soccer players but they did not realise that the sport could involve some mathematics: they erroneously thought that they would have to kick the ball high enough to clear it as far away as possible; obviously, it could not be at an angle of 90° from the ground but they believed it to be about 60° when in fact it should be 45°. Although this is a concept in physics, kinematics is also a branch of mathematics, not to mention that "mathematics is the queen of the sciences" (Reimer & Reimer, 1992, p. 83) — a famous quotation by the great mathematician Carl Gauss (1777-1855). Mathematics can also help soccer players to make a more informed decision if they know which position on the soccer field will give them the widest angle to shoot the ball between the goalposts (for more information, see Goos, Stillman & Vale, 2007, pp. 50-58).

However, there are still many jobs that may not require much mathematics, except perhaps for simple arithmetic like counting money and telling time, e.g., actors and actresses, taxi drivers, administrators, historians and language teachers. How often do they use, for example, algebra, in their workplace? Why study so much mathematics when many adults do not even use it in their professions? Isn't it a waste of time and resources?

I will address these issues by first examining the purposes of mathematics and the aims of mathematics education. Then I will discuss the rationales for a broad-based school curriculum which pertains not only to the learning of mathematics specifically but also the study of other subjects in general. After that, I will focus on real-life applications of mathematics in and outside the workplace by giving suitable examples that secondary school and junior college students can understand. But mathematics involves not just concepts and procedural skills but also thinking processes. Therefore, I will discuss how mathematical cognitive and metacognitive processes, such as investigation proficiency, problem-solving strategies, communication skills, and critical and creative thinking, are important in and outside the workplace in our daily life.

2 Purposes of Mathematics and Aims of Mathematics Education

In ancient times, mathematical practices were divided into scientific mathematics and subscientific mathematics (Høyrup, 1994): the former being mathematical knowledge pursued for its own sake without any intentions of applications; and the latter being specialists' knowledge that was applied in jobs such as architects, surveyors and accountants. Subscientific mathematics was often transmitted on the job and not found in books, but in modern days where mathematics is divided into pure and applied mathematical knowledge, the latter is usually taught at higher levels in schools because of the utilitarian ideology of the industrial trainers and the technological pragmatists who argue that schools should prepare students for the workforce (Ernest, 1991). However, groups with a purist ideology believe in studying mathematics for its own sake. For example, the old humanists advocate the transmission of mathematical knowledge because of their belief that mathematics is a body of structured pure knowledge, while the progressive educators prefer to use activities and play to engage students so as to facilitate self-realisation through mathematics because of their belief in the process view of mathematics and child-centreness. There is a third type of ideology called the social change ideology of the public educators who wish to create critical awareness and democratic citizenship via mathematics through the use of questioning, discussion, negotiation and decision making.

The different ideologies of these diverse social groups have affected and will continue to affect the aims of mathematics education (Cooper, 1985). For example, the New Math movement in the United States of America in the 1950s was an attempt by mathematicians, who could be viewed as old humanists with a purist ideology, to integrate the contents of academic mathematical practice into the school curriculum by teaching students more advanced mathematical concepts, such as commutative, associative and distributive laws, sets, functions and matrices, in order to understand the underlying algebraic structures and unifying concepts (Howson, 1983; Howson, Keitel & Kilpatrick, 1981). But many students were unable to cope with this type of high-level mathematics and they ended up not even skilled in basic mathematical concepts that were needed in the workforce. As a result, the industrial trainers and the technological pragmatists (groups with utilitarian ideology) won and it was 'Back to Basics' for school mathematics in the 1970s (Askey, 1999; Goldin, 2002).

Then, in the 1980s, the Standards of the National Council of Teachers of Mathematics (NCTM) (NCTM, 1980, 1989) sought to bring the *processes* of academic mathematics (as opposed to the *contents* of academic mathematics in the New Math movement) into the mathematics classrooms (Lampert, 1990; Schoenfeld, 1992). This model views students as mathematicians and it suggests that students should focus on activities that parallel what mathematicians do, for example, posing problems to solve, formulating and testing conjectures, constructing arguments and generalising (Moschkovich, 2002). This is more in line with the purist ideology of the progressive educators and perhaps even the social change ideology of the public educators who believe in the use of questioning and negotiation. It will be argued in Section 6 that these thinking processes have become important in the modern workplace where the requirements of the workforce are ever changing and technical skills are no longer enough (Wagner, 2008).

However, the needs of the society may differ from the desires of individual students: the latter may not be interested to learn mathematics partly because they do not see the use of it, especially when most applications in the workplace are beyond their level, and partly because they may not be interested in professions that require a lot of mathematics. Therefore, there is a need to convince these students why they still need to study mathematics (and other subjects) which they may not need in their future working life — the rationales for a broad-based school curriculum, before I proceed to give examples of real-life mathematical applications in our daily life.

3 Rationales for a Broad-Based School Curriculum

Many students have ambitions but how many of them will end up fulfilling them? Just because a student wants to be a doctor does not mean he or she has the ability or the opportunity to become one. In Singapore, there is a limited vacancy for studying medicine in the university because the government does not want an oversupply of doctors (which is a popular choice of career here) and an undersupply in other jobs since the only asset that Singapore has is its people as there are scarce natural resources. So, even if a student has the aptitude to become a doctor, he or she may still lose out to the more academically-inclined students, and unless the student's parents have the means to send him or her overseas to pursue his or her ambition, the student will most likely end up with another job. This is the first thing that teachers can impress upon their students.

The second thing is for teachers to ask their students whether the ambition that they have now is their first ambition. Many students would have changed their ambitions at least once. Even if this is still their first ambition, there is no guarantee they will not change their minds in future. So, what will happen if students are allowed to specialise early? If their interests are to change later, it will be too late to switch to another subject. Moreover, some students cannot decide at this moment what job they want to do in future, so it is impossible to force them to choose now. Therefore, this is the basis of a broad-based school curriculum, to teach students the basics of different subjects so that when they need it in their area of specialisation later on, they will have some foundations to build upon.

Another reason for a broad-based curriculum is to expose students to different subjects so that they can find out where their interests lie. For example, if a student has never studied geography, how will the student know whether he or she will like to be a geologist or not? But what happens if a student studies a subject and then decides this is not for him or her in his or her future working life? Does that mean that the student can give up that subject? Although teachers can reason with their students that they may end up with a job involving a subject that they dislike in school, it may be more worthwhile to show them how they can still use the subject outside the workplace or in a different area of work, e.g., mathematical thinking processes can be applied to solve unfamiliar problems in other fields.

In the rest of this chapter, I will provide specific examples of how mathematics is applicable in our daily life. I distinguish between mathematical knowledge and processes, and between mathematics in and outside the workplace. Some educators (e.g., Moschkovich, 2002) have classified workplace mathematics as a subset of everyday mathematics.

4 Applications of Mathematical Knowledge in the Workplace

What are in the Singapore secondary syllabus and textbooks are mostly arithmetical applications such as profit and loss, discount, commission, interest rates, hire purchase, money exchange and taxation. But what about workplace uses of algebra, geometry, trigonometry and calculus? Usually, many of these applications are beyond the level of most students. However, this section will illustrate some suitable real-life applications which teachers can discuss with their students.

4.1 *Use of similar triangles in radiation oncology*

Geometry plays a very important role in radiation oncology (the study and treatment of tumours) when determining safe level of radiation to be administered to spinal cords of cancer patients (WGBH Educational Foundation, 2002). Figure 1 shows how far apart two beams of radiation must be placed so that they will not overlap at the spinal cord, or else a double dose of radiation will endanger the patient.

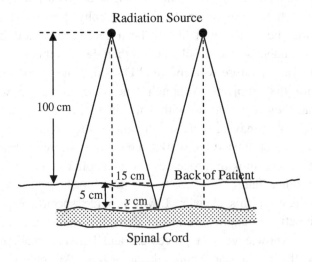

Figure 1. Radiation therapy for cancer patients

In this case, the distance of the radiation source from the back of the patient is 100 cm and 15 cm is only the length on the back. The actual length on the spinal cord, x cm, can be determined using a property of similar triangles:

$$\frac{100}{105} = \frac{15}{x}$$

$$x = \frac{15 \times 105}{100}$$

$$= 15.75$$

Therefore, the actual length is longer by 0.75 cm. If the two beams of radiation are placed only 2×15 cm = 30 cm (instead of 2×15.75 cm = 31.5 cm) apart, then 2×0.75 cm = 1.5 cm of the spinal cord will be exposed to a double dose of radiation.

4.2 Optimisation problems using calculus

Another real-life application is to minimise the total surface area S of a cylindrical can food for a *fixed* volume V in order to cut cost for the metal used. Since $V = \pi r^2 h$, where r is the radius and h is the height of the cylinder, then $h = \dfrac{V}{\pi r^2}$. Therefore, $S = 2\pi r h + 2\pi r^2 = \dfrac{2V}{r} + 2\pi r^2$.

Using the first derivative test, we can find the value of r that gives the least value of S. What is interesting is the relationship between h and r when S is least: $h = 2r$, i.e., the height of the can is equal to its diameter. But many cylindrical can foods are not of these dimensions. Is it because the manufacturer does not know calculus or are there other factors? This is what happens in real life: there is more to it than mathematics. Teachers can ask their students why this is so. Below are some reasons:

- Hotdogs and sardines are long, so their cans are usually longer in shape, i.e., their height is greater than their diameter.
- Can drinks are also longer in shape because it is harder for our hands to hold them if their diameter is equal to their height.

- A can, whose diameter is equal to their height, may be aesthetically less appealing to the consumers.
- The metal sheets may come in a particular size such that fewer cans with equal diameter and height can be made from one metal sheet than if the cans are of a different dimension.

An article on John Handley High School (2006) Website reveals some more reasons why manufacturers do not make cylindrical cans which have equal height and diameter:

- Most can manufacturing lines run a variety of can sizes. To change both the height and diameter for cans with a different volume will require more time than just changing the height and using the same ends.
- Tall thin cans or short wide cans require less thermal processing time to sterilise than cans with equal height and diameter.
- Internal pressure that develops during thermal processing can distort the ends of the can, so the ends are made of thicker metal which is more costly. But if the ends are smaller in diameter, thinner metal can be used for both the ends and curved surface.
- Tall thin cans make better use of packaging and shipping space.

The last three reasons reveal that there are other types of mathematics involved, e.g., in thermal processing, and in packaging and shipping space.

4.3 *Encoding of computer data using large prime numbers*

Before computer data are sent through the Internet, especially sensitive data like online banking, they are encoded so that hackers cannot make use of the data unless they know how to decode. Ronald **R**ivest, Adi **S**hamir and Leonard **A**dleman have invented the very secure RSA cryptosystem which makes use of very large primes and a complicated number theory. To break the code, hackers need to decompose a very large composite number into its two very big prime factors using trial division. Even with the fastest computers, this will take years because trial division for very large numbers is an extremely tedious process. Although the actual encryption (refer to Wolfram MathWorld Website, Weisstein, 2009a, for more information) is beyond the level of the students, there are three other things that teachers can do with their students.

The first thing to discuss is how big these primes are. Teachers can introduce to their students the search for very large Mersenne primes. Mersenne (1588-1648) used the formula $M_p = 2^p - 1$, where p is prime, to generate big primes, but M_p is only prime for certain values of p, e.g., M_2, M_3, M_5 and M_7 are primes but M_{11} is not a prime. The largest known prime, $M_{43,112,609}$, found by Edson Smith on 23 Aug 2008, contains 12 978 189 digits! If one newspaper page can contain 30 000 digits, then 433 newspaper pages will be needed to print the largest known prime! For more information, refer to the Great Internet Mersenne Prime Search or GIMPS Website (Mersenne Research, 2009).

Another interesting issue to discuss is the complicated number theory that is the basis of the RSA cryptosystem. For centuries, mathematicians had studied prime numbers and number theory for interest sake since there were no known real-life applications at that time. But because of their study, with the invention of computers, there is now a very secure way of protecting sensitive computer data. Teachers can let their students debate the tension and merit between education for interest sake and education to prepare students for the workplace — the purist ideology versus the utilitarian ideology discussed in Section 2.

The third thing is to let students have a sense of encryption by using some other simpler but not-so-secure systems since they will not be able to understand the RSA cryptosystem. An example of such a system is the use of a matrix to encode and its inverse to decode the data. The reader can refer to Teh, Loh, Yeo and Chow (2007a) pp. 44-51 for more details and a ready-to-use student worksheet.

4.4 *Use of differentiation in economics*

Economics, a subject which some junior college students take, uses differentiation extensively. For example, retailers are interested to find out whether an increase in the price of a product will cause the demand to drop so drastically that the total revenue will fall? Or will the demand drop only slightly such that the total revenue will still increase? In economics, this is called the price elasticity of demand and it measures the degree of responsiveness in the quantity demanded of a commodity as a result of a change in its price (Wikipedia, 2009a). Suppose a retailer does a survey and finds out that the daily demand curve of a particular product is given by $Q = 20 - 0.5P$, where Q is the quantity demanded of the commodity and P is its price in dollars. The price elasticity of demand E_d is calculated as follows:

$$E_d = \frac{dQ}{dP} \times \frac{P}{Q} = -0.5 \times \frac{P}{Q}.$$

What does this mean? If $|E_d| = 1$ (unit elastic), the percentage change in quantity demanded is equal to the percentage change in price. If $|E_d| > 1$ (elastic), the percentage change in quantity demanded is bigger than that in price, and so if the price is raised, the quantity demand will drop enough to cause the total revenue to fall, and if the price is lowered, the quantity demand will rise enough to cause the total revenue to increase. If $|E_d| < 1$ (inelastic), the percentage change in quantity demanded is smaller than that in price, and so if the price is increased, the total revenue will still rise despite the drop in demand, and vice versa.

Suppose the retailer is selling the product for $30, i.e., $P = 30$. Then $Q = 20 - 0.5P = 5$. Since $|E_d| = |-0.5 \times 30 \div 5| = 3 > 1$, then the retailer should not increase but decrease the price. For example, if the retailer increases the price by $2 from $30 to $32, then $Q = 4$ and the total revenue becomes $P \times Q = \$32 \times 4 = \$128$, which is a lot less than the original total revenue of $30 \times 5 = \$150$. But if the retailer decreases the price by $2 to $28, then $Q = 6$ and the total revenue becomes $168, which is more than the original total revenue of $150. Figure 2 shows the graph of Q against P and the graph of the total revenue TR against P.

The question is how low the price can drop before the total revenue starts to decrease. Unit elasticity, i.e., $|E_d| = 1$, occurs when $P = 20$ (and $Q = 10$), so the retailer should lower the price to $20 for the total revenue to reach its peak of $20 \times 10 = \$200$ (refer to the maximum point of the bottom graph in Fig. 2). If the price is decreased further to, say, $18, then $Q = 11$ and the total revenue is only $18 \times 11 = \$198 < \$200$.

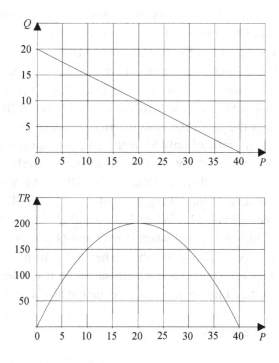

Figure 2. Graphs of Q against P and TR against P

In fact, when $P = 18$ and $|E_d| \approx 0.818 < 1$, i.e., the price elasticity of demand is inelastic, the retailer can increase or decrease the price without affecting the demand so much. This means the retailer should raise the price so that the total revenue will increase. But how high can the price rise before the total revenue starts to decrease? The answer is still the same: when unit elasticity is reached, i.e., when the price is $20 and the total revenue is at its maximum of $200.

In real life, things are not that simple. For example, how does one conduct a survey to obtain a reliable demand curve? And if the retailer drops the price from $30 to $20, he or she will have to bring in more stock, which translates to more storage space, more capital if the supplier refuses to give the retailer a higher credit, more time and effort in bringing in and selling more goods, and a higher risk if the product does not sell according to the demand curve.

A question some students may ask is whether it is possible for the total revenue to fall but the profit will actually increase? For example, if the retailer makes a profit of $1 for each item, when $P = 30$ and $Q = 5$, he or she only earns $5. But if the price is increased by $2 to $32, then $Q = 4$ and the total revenue falls from $150 to $128, but the profit increases from $5 to $3 × 4 = $12. This scenario can only happen if the profit per item is lower than the small increase in price. However, in real life, no retailer will make a profit of only $1 if the cost price of the item is $29 because a 3.4% profit will not be enough to cover overheads such as rental and salaries of employees. Retailers usually make about 40% to 60% profit per item, so if the selling price of the item is $30, the retailer will make between $9 and $11. When $P = 30$ and $Q = 5$, the retailer will make a profit of $45 to $55, but when $P = 32$ and $Q = 4$, the retailer will earn about $44 to $52. Therefore, the higher the percentage profit per item, a small increase in price and a corresponding decrease in demand will not only result in a fall in the total revenue but also a drop in profit.

4.5 *Other generic examples*

It is usually not possible to find a real-life application for every topic that students can understand. But there are applications that teachers can still discuss with their students without going into specific details. For example, complex numbers are used extensively in electrical engineering to understand and analyse alternating signals (United States Naval Academy Website, 2001); GPS (global positioning system) makes use of complex vectors and geometric trilateration to determine the positions of the objects (Wikipedia, 2009b); and land surveying equipment uses trigonometry and triangulation (Wikipedia, 2009c). Teachers can also let their students have a sense of how the latter works by using a clinometer (a simple instrument that measures the angle of elevation), a measuring tape and trigonometry to find the height of a tree or a building (for more details and a ready-to-use worksheet, see Teh, Loh, Yeo & Chow, 2007a, pp. 108-111).

Another example is the use of the formulae for finding arc length and sector area, and the symmetric and angle properties of circles, in the design and building of road tunnels, bridges, buildings and any arc-shaped structures. Teachers can let their students have a sense of how this works by getting them to draw an arc-shaped balcony, given the following dimensions in Figure 3. To construct the arc, the students will have to find the centre and radius of the corresponding circle using some circle properties. In fact, there is more than one way to do it.

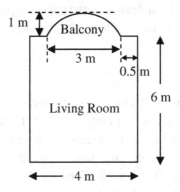

Figure 3. Arc-shaped balcony

An inspiring story to tell is the ingenious plug-in called Auto-Tune that can transform mediocre singing into note- and pitch-perfect singing (Tyrangiel, 2009). Many students are interested in songs and so using an example on singing may capture their attention. The inventor is Andy Hildebrand who has worked for many years in the oil industry by providing an accurate map of potential drilling sites using sound waves sent into the ground and recording their reflections. This technique, which uses a mathematical formula called autocorrelation (refer to Wolfram MathWorld Website, Weisstein, 2009b, for more information), has saved oil companies lots of money and allows Hildebrand to retire at 40! After retirement, at a challenge from a dinner party guest to invent something that would allow her to sing in tune, he invented Auto-Tune, which also uses autocorrelation. Even if students do not know how autocorrelation works, they may be impressed that a mathematical formula can allow a person to earn so much money that he can retire at 40! This shows the usefulness of mathematics in the workplace.

5 Applications of Mathematical Knowledge Outside the Workplace

Although mathematics is useful in many professions, what if students end up with occupations that do not require much mathematics? Does that mean the mathematics education that they have received is all gone to waste? Therefore, I will discuss in this section how mathematics is useful outside the workplace in everyday life, e.g., understanding everyday events and analysing newspaper reports, even when the students have become adults holding jobs that do not need mathematics.

5.1 *Earthquakes*

Logarithm is used in the Richter scale to measure the magnitude of an earthquake, e.g., the Indonesian earthquake, that set off giant tidal waves on 26 Dec 2004 and killed at least 159 000 people, measured 9.0. What does this magnitude mean and how does it compare with a 4.8-magnitude earthquake that struck Northern Italy on 5 Apr 2009 and killed almost 300 and left 60 000 homeless? Although some people may know that a 9.0-magnitude earthquake is ten times stronger than an 8.0 earthquake, complications arise when the difference in magnitude is not an integer.

The Richter scale is a measurement of earthquake magnitudes based on the formula $R = \lg\left(\dfrac{x}{0.001}\right)$ where x is the intensity of the earthquake as registered on a seismograph. Teachers can get their students to find x in terms of R by transforming the logarithmic form to the index form (this procedural skill is tested in GCE O-level exam!), i.e., $x = 0.001 \times 10^R$. To compare two intensities x_2 and x_1, we have:

$$\frac{x_2}{x_1} = \frac{0.001 \times 10^{R_2}}{0.001 \times 10^{R_1}} = 10^{R_2 - R_1}.$$

Thus, to contrast the strength of two earthquakes, we just need to find the difference in magnitude $d = R_2 - R_1$ and then calculate 10^d. For example, the difference in magnitude between the Indonesian earthquake that set off the tsunami, and the recent earthquake in Northern Italy, is

$d = 9.0 - 4.8 = 4.2$, and so the Indonesian earthquake is $10^{4.2} \approx 15\,800$ times stronger than the Italian one. Although the difference of 4.2 in magnitude looks small, the difference in intensity is actually a lot bigger, which has surprised many students. For a ready-to-use student worksheet on earthquakes and logarithms, the reader can refer to Teh, Loh, Yeo and Chow (2007b) pp. 25-28.

5.2 *Lottery*

A 4D draw in Singapore has a total of 23 prizes: first, second and third prizes, ten starter prizes and ten consolation prizes. There are 10 000 possible four-digit numbers: 0000 to 9999. The first and second prizes for the 4D Draw on 27 Jun 2007 were both the same number 6904. The Sunday Times on 8 Jul 2007 quoted a spokesman from Singapore Pools (the company that organises the 4D Draw) as saying that the probability for a number to appear twice in the same draw is 1 in 10 000 times 10 000, or one in 100 millions (Mak, 2007). Surprisingly, many mathematics teachers, whom I have spoken to, agreed with the statement.

Let us begin with a simpler question: what is the probability for the same number to appear as first and second prizes in the same draw? Consider a related problem: what is the probability of obtaining the same number in a toss of two dice? The total number of possible outcomes for throwing two dice is 36. The favourable outcomes are (1, 1), (2, 2), (3, 3), (4, 4), (5, 5) and (6, 6). Thus the probability of obtaining the same number on the two dice is $\frac{6}{36} = \frac{1}{6}$. Similarly, the total number of possible outcomes for the first and second prizes is 10 000 × 10 000. The favourable outcomes are (0000, 0000), (0001, 0001), (0002, 0002), ..., (9999, 9999) and so the probability of obtaining the same number for the first two prizes is $\frac{10\,000}{10\,000 \times 10\,000} = \frac{1}{10\,000}$.

Another method is as follows. Since it does not matter what the number on the first die is, the probability of obtaining any number on the die is 1. Then there is only one favourable outcome for the second die and so the probability of obtaining the same number on both dice is

$1 \times \frac{1}{6} = \frac{1}{6}$, and not $\frac{1}{6} \times \frac{1}{6} = \frac{1}{36}$. Using the same argument, the probability of the same number appearing as the first and second prizes in the same draw is $1 \times \frac{1}{10\,000} = \frac{1}{10\,000}$, and not $\frac{1}{10\,000} \times \frac{1}{10\,000} = \frac{1}{100\,000\,000}$.

Quite a number of mathematics teachers, whom I have spoken to, were still not convinced. The main confusion is that it now *seems* that the probability of a number appearing twice in a draw is equal to the probability of a draw having all different numbers, which the teachers have this intuitive feeling that it is not right. But the probability of a number appearing twice in a draw is *not* $\frac{1}{10\,000}$; rather, $\frac{1}{10\,000}$ is the probability of a number appearing in both the first and second prizes. Similarly, the probability of a draw having all different numbers is not $\frac{1}{10\,000}$, but it is equal to $\frac{^{10\,000}P_{23}}{10\,000^{23}} \approx 0.975$ since there are 23 prizes in a draw. Secondary school students, who have not learnt permutations, can find the probability as follows, although computing it is rather tedious:

$$1 \times \frac{9999}{10\,000} \times \frac{9998}{10\,000} \times \frac{9997}{10\,000} \times \ldots \times \frac{9978}{10\,000}.$$

Thus the probability of a draw having at least one number appearing more than once is about 0.025. As there are three draws every week, and so about 156 draws a year, we should expect about four draws to have at least one number appearing more than once every year. Although the latter is rarer than a draw having all different numbers, it is theoretically not so rare at all! However, in reality, most years have gone by without a single draw having at least one number appearing more than once.

5.3 *Water supply*

An article from Streats (a free newspaper in Singapore that has since ceased publication) on 26 May 2001 quoted Mr Lim Swee Say, then

Acting Minister for the Environment of Singapore, as saying, "If we increase the supply of fresh water by 20 per cent and at the same time reclaim 30 per cent of the used water, we will be able to increase our total water capacity by as much as 70 per cent." ("Water: Add and Multiply", 2001) But 20% plus 30% of 1.2 is only 56%. Was Mr Lim wrong?

Let V be the volume of water that we have. If we increase the supply of fresh water by 20%, then the total volume of water will be $1.2V$. If we reclaim 30% of the used water (all the $1.2V$ of water will be used after some time), then the volume of reclaimed water will be $0.3 \times 1.2V$, i.e., the total volume of water will be $1.2V + 0.3 \times 1.2V = 1.56V$, an increase of 56%. But this is only after the first reclamation. The reclaimed water will be used after some time and then we can still reclaim 30% of it. So the total volume of water after the second reclamation will be $1.2V + 0.3 \times 1.2V + 0.3^2 \times 1.2V$. Continuing this process of reclamation, the total volume of water in the long run will be:

$$1.2V + 0.3 \times 1.2V + 0.3^2 \times 1.2V + 0.3^3 \times 1.2V + \ldots$$
$$= 1.2V \times (1 + 0.3 + 0.3^2 + 0.3^3 + \ldots)$$
$$= 1.2V \times \frac{1}{1 - 0.3}$$
$$\approx 1.71V$$

Therefore, Mr Lim was correct to say that it was possible to increase our water supply by as much as 70%. A question that teachers should invite their students to ask is, "Do we have to wait forever to get a 70% increase since this is the sum to infinity? Or after how many times of reclamation will we get near enough to a 70% increase?" Fortunately, we do not have to wait so long: after the third reclamation, the total volume of water will be $1.7004V$. Although the initial calculation involves summing to infinity for a geometric progression (GP), which is only taught at the junior college level, secondary school students can actually find the answer for three times of reclamation without using the GP formula.

5.4 Quality of life

Some students may question the purpose of understanding everyday events and analysing newspaper reports as outlined above. The American philosopher and educational theorist John Dewey (1859–1952) believed that education was a process of enhancing the quality of life. "By *quality*, Dewey meant a life of meaningful activity, of thoughtful conduct, and of open communication and interaction with other people." (Hansen, 2007, p. 9) If people are ignorant or cannot make sense of the events that happen during the daily course of their lives, then they may not be able to engage in meaningful discourse with other people, make well-informed decision and lead a life of accountable conduct.

However, some people think that life will still go on if they do not understand all these. But ignorance is not bliss. They may make bad ill-informed decisions because they do not understand, for example, the spread of the H1N1 flu virus as reported in the media. Some elderly people think that they are immune to the virus since it attacks mostly young adults with medical problems (personal communication), and so the former may not take any precautions; this not only affects themselves but they may end up spreading the virus to others. Hence, knowledge and the ability to reason logically are necessary not only for leading a meaningful life but also a responsible one.

6 Applications of Mathematical Processes in the Workplace

Mathematics is not only about conceptual knowledge and procedural skills, but it involves cognitive processes such as problem-solving heuristics and thinking skills; metacognitive processes such as the monitoring and regulation of one's own thinking; and attitudes such as interest, confidence and perseverance in solving unfamiliar problems. Concepts, skills, processes, metacognition and attitudes are the five components of the Pentagon Model which is the framework for the Singapore mathematics curriculum (Ministry of Education, 2006). But are these thinking processes, metacognition and affective variables important in the workplace, and if yes, why?

Tony Wagner, Co-director of the Change Leadership Group at the Harvard Graduate School of Education, interviewed CEOs of big companies to find out what key skills they valued most when hiring new staff (Wagner, 2008). To his surprise, many of them put critical thinking and problem solving as one of the top priorities and "the heart of critical thinking and problem solving is the ability to ask the right questions" (p. 21). For example, Clay Parker, President of the Chemical Management Division of BOC Edwards, a company that makes machines and supplies chemicals for the manufacture of microelectronics devices, replied that he would look for someone who asked good questions. He said, "We can teach them the technical stuff, but we can't teach them how to ask good questions — how to think." (p. 20) (Yet, teachers are expected to teach their students how to think!) As one senior executive from Dell revealed, "Yesterday's answers won't solve today's problems." (p. 21)

Other vital skills include leadership and collaboration, agility and adaptability, initiative and entrepreneurialism, effective oral and written communication, accessing and analysing information, and curiosity and imagination. Thus there is value in developing students' mathematical problem-solving strategies, investigative and research skills, analytical, critical and creative thinking processes, teamwork and communication, and arousing their curiosity and interest in mathematics, because these are essential skills in the workplace. Although the domain may not be mathematics per se, hopefully, students are able to transfer their habits of minds to new and unfamiliar situations in other fields. If we believe that one of the purposes of education is to prepare students for the workforce (Cowen, 2007), then schools and teachers should heed Wagner's (2008) advice to teach and test the skills that matter most.

7 Applications of Mathematical Processes Outside the Workplace

Mathematical processes are also important outside the workplace in everyday life, e.g., analysing statistical data reported in newspapers critically so as not to be deceived by biased reporting (in fact, this type of analytical skills is also important in the workplace when reading reports from clients). In this section, we will analyse how statistics, which is the science of collecting, organising, displaying and interpreting data (Yap,

2008), can be manipulated in each of the four components to tell different stories, whether intentionally or unintentionally.

7.1 *Collection of data*

In a UEFA (Union of European Football Associations) Website poll involving 150 000 fans, Zinedine Zidane was named the best European footballer in the past 50 years (Reuters, 2004). He polled 123 582 votes, while second-placed Franz Beckenbauer received 122 569 votes, followed by Johan Cruyff with 119 332 votes. Teachers can ask their students this question: "Do most fans *really* think that Zidane is the best European soccer player of the last 50 years?" Most students, who are soccer fans, know who Zidane is, but few have heard of Beckenbauer or Cruyff. This is because Beckenbauer and Cruyff were famous in the 1970s while Zidane was at the peak of his career in the 1990s. Moreover, the poll was conducted online, meaning that the older fans who knew Beckenbauer and Cruyff were less likely to vote since they are usually less technological savvy than the younger fans. Despite all these, the polls show that Beckenbauer was only slightly behind Zidane — about 1000 votes or less than 1%. Therefore, if the polls were conducted in such a way that all the people who voted knew who Beckenbauer was, most likely Beckenbauer would dethrone Zidane. This is an example of how an inappropriate polling method can affect the voting outcome: the convenient sample is not representative of the population.

7.2 *Organisation of data*

The first two paragraphs of a Straits Times article *Insurance firms, banks, top hate list* read, "Banks and insurance companies have made it to the top of the consumer hate list for the first time. They were the target of 1,915 complaints to the Consumers Association of Singapore (Case) between January and last month, edging out the usual suspects — timeshare companies (1,228 complaints), motor vehicle shops and companies (1,027), renovation companies (963), and electrical and electronics shops (710)." (Arshad, 2003, p. 1) Teachers can get their students to question the report critically. One of the main problems of this article was the organisation of the data: why were banks and insurance companies grouped together, while the others were single entities? Although banks do sell insurance, they are still different from insurance companies. If banks and insurance companies were considered separately, it might be possible that timeshare companies would now top the hate list, e.g., if the 1915 complaints were split equally between banks and insurance companies. This is an example of how the decision to group or not to group the data, and how the method of grouping, can distort the truth.

7.3 *Display of data*

The headline of a survey reported on the front page of the Straits Times was SEX AND THE SINGAPORE GIRL (Mathi, 1998). It showed a big picture of two students kissing. The subheading 'Girls date — and kiss — before boys do' was supported by the first two bar graphs in the article, and the fourth bar graph showed that more girls had engaged in sex than boys. Figure 4 reproduces these three bar graphs in the article.

On page 26 of the same newspaper, a small report (with no picture) of the same survey stated, "Overall, the ST survey gave a reassuring picture of the Singapore teen. As Dr V. Atputharajah, Kandang Kerbau Women's and Children's Hospital's senior consultant with the department of reproductive medicine, said: 'Many teenagers have this impression that everyone is having sex. But in reality this is not true, *as the survey has shown* [italics mine].'" (Mathi & Wong, 1998, p. 26) But most readers, including my students at that time, had the impression that many girls were sexually active because of the front page news with such a telling photograph since a picture says a thousand words.

If the bar graphs were displayed differently as shown in Figure 5, then the picture presented will be entirely different: most of the girls did not date at 13 years or younger (85%), did not kiss by 14 years or younger (92%) and did not ever have sex (93%). This agrees with what Dr Atputharajah has said on page 26, "Many teenagers have this impression that everyone is having sex. But in reality this is not true, as the survey has shown." Hence, the display of data can affect how the readers perceive the survey results.

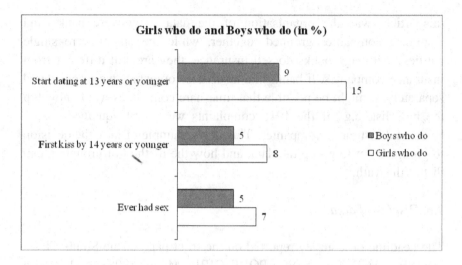

Figure 4. Bar graphs comparing girls who do and boys who do
(as depicted in the Strait Times article)

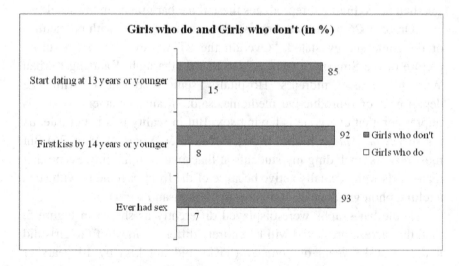

Figure 5. Bar graphs comparing girls who do and girls who do not
(which tell a different story)

7.4 Interpretation of data

The headline of a survey reported in the Straits Times was 'Survey finds discipline in schools not a big problem' (Mathi, 1997). The survey interviewed 285 teachers over two years and three out of five teachers said that the discipline problem was not serious and so the researcher concluded, "This is the good news — that the state of discipline in schools is not as bad as it has been made out to be." Of course, one can question about the sampling, but the main focus in this section is the interpretation of data. In order for discipline to be viewed as a serious problem, does one need more than 50% of teachers to agree that it is? Think about this: in a school with 100 teachers, if 40 teachers believe that the discipline problem in the school is bad enough, do *you* just treat it lightly and dismiss it? Most of the teachers, whom I have spoken to, agreed that it should be a concern if 40% of the teachers believed that the discipline in their school was serious.

The key issue here is that one does not always need a simple majority (i.e., more than 50%) to decide on a matter. For example, to pass a law in the Singapore Parliament that results in a constitutional amendment, at least two-thirds of the elected Members of Parliament must agree on it; a simple majority is not enough. On the other hand, a simple majority may not even be needed for an issue to become a problem in some cases. This is an example of how the basis or criterion for making a decision can affect the interpretation of the data.

7.5 Can statistics lie?

Evan Esar (1899-1995) believed that statistics was the only science that enabled different experts using the same figures to draw different conclusions (Moncur, 2007). But we go one step further: even the figures can lie! If teachers do not teach their students to analyse survey reports critically, then they may be misled by others. Although it is good to make informed decisions in everyday lives, it is even more important in the workplace because a mistake in judging the reliability of reports from clients can cause the company to suffer huge financial loss. However, the con of teaching students how to evaluate the validity of the collection, organisation, display and interpretation of data is that the students may end up using this knowledge to mislead others. That is why moral education in school is also very important: an intelligent crook is worse than a stupid one! Teachers should also teach their students to be responsible and useful citizens — the social change ideology of the public educators (Ernest, 1991) discussed in Section 2.

8 Conclusion

Mathematics is of practical value in many professions. It is not just the mathematical knowledge itself but the thinking processes acquired in genuine mathematical problem solving and investigation that can be applied to unfamiliar situations in other fields. Mathematical knowledge and processes are also useful outside the workplace in everyday life to understand and interpret certain events and news reports so as not to be deceived or swayed by others' opinions without any reasonable basis, thus improving one's own quality of life when one is able to lead a meaningful and responsible life. Teachers should impress upon their students the usefulness of mathematics in their daily life, and they should prepare their students for the future by focusing on the essential skills and processes that are required in the workplace.

References

Arshad, A. (2003, December 21). Insurance forms, banks top hate list. *The Straits Times*, p. 1.
Askey, R. (1999). The third mathematics education revolution. In E. A. Gavosto, S. G. Krantz, & W. McCallum (Eds.), *Contemporary issues in mathematics education* (pp. 95-107). Australia: Cambridge University Press.

Cooper, B. (1985). *Renegotiating secondary school mathematics: A study of curriculum change and stability*. London: Falmer Press.

Cowen, R. (2007). Last past the post: Comparative education, modernity and perhaps post-modernity. In M. Crossley, P. Broadfoot, & M. Schweisfurth (Eds.), *Changing educational contexts, issues and identities: 40 years of comparative education* (pp. 148-174). London: Routledge.

Ernest, P. (1991). *The philosophy of mathematics education*. London: Falmer Press.

Goldin, G. A. (2002). Representation in mathematical learning and problem solving. In L. D. English (Ed.), *Handbook of international research in mathematics education* (pp. 197-218). Mahwah, NJ: Erlbaum.

Goos, M., Stillman, G., & Vale, C. (2007). *Teaching secondary school mathematics: Research and practice for the 21st century*. NSW, Australia: Allen & Unwin.

Hansen, D. T. (Ed.). (2007). *Ethical visions of education: Philosophies in practice*. New York: Teachers College Press.

Howson, A. G. (1983). *A review of research in mathematics education: Part C. Curricular development and curricular research. A historical and comparative view*. Windsor, Berks: NFER-Nelson.

Howson, G., Keitel, C., & Kilpatrick, J. (1981). *Curriculum development in mathematics*. Cambridge: Cambridge University Press.

Høyrup, J. (1994). *In measure, number, and weight: Studies in mathematics and culture*. Albany: SUNY Press.

John Handley High School (2006). *Cat food can problem*. Retrieved July 16, 2009, from www.doe.virginia.gov/Div/Winchester/jhhs/math/lessons/calculus/catfood.html

Lampert, M. (1990). When the problem is not the question and the solution is not the answer: Mathematical knowing and teaching. *American Educational Research Journal, 27*, 29-63.

Mak, M. S. (2007, July 8). 4-D a winner? You bet. *The Sunday Times*, p. L4.

Mathi, B. (1997, December 1). Survey finds discipline in schools not a big problem. *The Straits Times*, page number not available.

Mathi, B. (1998, March 24). Sex and the Singapore girl. *The Straits Times*, p. 1.

Mathi, B., & Wong, C. M. (1998, March 24). They don't go on dates. *The Straits Times*, p. 26.

Mersenne Research. (2009). *Great Internet Mersenne Prime Search (GIMPS)*. Retrieved July 16, 2009, from www.mersenne.org

Ministry of Education. (2006). *Mathematics syllabus: Secondary*. Singapore: Author.

Moncur, M. (2007). *The quotations page*. Retrieved July 16, 2009, from www.quotationspage.com

Moschkovich, J. N. (2002). An introduction to examining everyday and academic mathematical practices. In E. Yackel (Series Ed.) & M. E. Brenner & J. N. Moschkovich (Monograph Eds.), *Everyday and academic mathematics in the classroom* (pp. 1-11). Reston, VA: National Council of Teachers of Mathematics.

National Council of Teachers of Mathematics. (1980). *An agenda for action: Recommendations for school mathematics of the 1980s.* Reston, VA: Author.

National Council of Teachers of Mathematics. (1989). *Curriculum and evaluation standards for school mathematics.* Reston, VA: Author.

Reimer, W., & Reimer, L. (1992). *Historical connections in mathematics: Resources for using mathematics in the classroom* (Vol. 1). Fresno, CA: AIMS Educational Foundation.

Reuters. (2004, April 24). Zidane voted Europe's best in past 50 years. *The Straits Times,* page number not available.

Schoenfeld, A. H. (1992). Learning to think mathematically: Problem solving, metacognition, and sense making in mathematics. In D. A. Grouws (Ed.), *Handbook of research on mathematics teaching and learning* (pp. 334-370). Reston, VA: National Council of Teachers of Mathematics & Macmillan.

Teh, K. S., Loh, C. Y., Yeo, J. B. W., & Chow, I. (2007a). *New Syllabus Mathematics Workbook 3: Alternative Assessment & CD Included* (Rev. ed.). Singapore: Shinglee.

Teh, K. S., Loh, C. Y., Yeo, J. B. W., & Chow, I. (2007b). *New Syllabus Additional Mathematics Workbook: Alternative Assessment & CD Included* (Rev. ed.). Singapore: Shinglee.

Tyrangiel, J. (2009, February 16). Singer's little helper. *Time, 173*(6), 43-45.

United States Naval Academy. (2001). *Complex number tutorials.* Retrieved July 16, 2009, from www.usna.edu/MathDept/CDP/ComplexNum/Module_1/intro.htm

Wagner, T. (2008). Rigor redefined. *Educational Leadership, 66*(2), 20-24.

Water: Add and multiply. (2001, May 26). *Streats,* page number not available.

Weisstein, E. W. (2009a). RSA encryption. From *MathWorld* – A Wolfram Web Resource. Retrieved July 16, 2009, from http://mathworld.wolfram.com/RSA Encryption.html

Weisstein, E. W. (2009b). Autocorrelation. From *MathWorld* – A Wolfram Web Resource. Retrieved July 16, 2009, from http://mathworld.wolfram.com/Auto correlation.html

WGBH Educational Foundation (2002). *Application of geometry in radiation oncology.* Retrieved July 16, 2009, from mms://media.scctv.net/annenberg/lm_geo_08_b2.wmv

Wikipedia. (2009a). Price elasticity of demand. From *Wikipedia* – The Free Encyclopedia. Retrieved July 16, 2009, from http://en.wikipedia.org/wiki/Price_elasticity_of_demand

Wikipedia. (2009b). Global positioning system. From *Wikipedia* – The Free Encyclopedia. Retrieved July 16, 2009, from http://en.wikipedia.org/wiki/GPS

Wikipedia. (2009c). Triangulation. From *Wikipedia* – The Free Encyclopedia. Retrieved July 16, 2009, from http://en.wikipedia.org/wiki/Triangulation

Yap, S. F. (2008). Teaching of statistics. In P. Y. Lee (Ed.), *Teaching secondary school mathematics: A resource book* (2nd ed., pp. 87-104). Singapore: McGraw-Hill Education.

Chapter 10

Using ICT in Applications of Secondary School Mathematics

Barry KISSANE

Apart from easing some of the computational burden, an attraction of ICT for mathematics education is that we may have new opportunities for pupils to learn and teachers to teach. This chapter describes some recent examples of technologies that seem likely to be appropriate for use in Singapore, and analyses some of their potential for mathematical applications and modelling, especially in probability and statistics. Particular attention is paid to hand-held calculators because they are the most likely to be affordable and available, especially when the constraints of examinations are taken into account. Additionally, some computer software with particular strengths is acknowledged: spreadsheets, which are likely to be widely available where computers are available; *Fathom* and *Tinkerplots*, innovative commercial software for statistics. The Internet offers considerable potential for mathematics education, despite relatively high costs of obtaining access to it, and seems likely to grow in significance in Singapore in the next several years. Some ways in which the Internet might support increased attention to applications and modelling are identified. The chapter concludes with recognition of the pervasive significance of professional development of teachers.

1 Introduction

There have been remarkable changes in the past decade to the Information and Communications Technology (ICT) available to pupils, teachers and schools. These changes together suggest that a 21st century mathematics curriculum should be substantially different from curricula of earlier times. In this chapter, we review some of the changed possibilities for both pupils and teachers, and look particularly at the implications of these for applications of mathematics.

Applications of mathematics can be supported by ICT in a range of ways, including representing real-world situations, collecting and analysing data, representing mathematical models and undertaking computation. This chapter provides examples of some of the ways in which ICT use by pupils can contribute to their understanding of and competence in applying mathematics to real world situations, especially in the areas of probability and statistics.

In describing the aims of mathematics education in Singapore schools, the Ministry of Education (2006) has referred explicitly to the use of ICT and applications:

Mathematics education aims to enable students to:

Make effective use of a variety of mathematical tools (including information and communication technology tools) in the learning and application of mathematics. (p. 3)

As well as having some intrinsic importance, applications and modelling play a vital role in the development of mathematical understanding and competencies, as reflected in the Singapore secondary school syllabuses. It is important that pupils apply mathematical problem-solving skills and reasoning skills to tackle a variety of problems, including real-world problems. Their learning of mathematics and their understanding of its relevance to modelling will be supported by using mathematical ideas and thinking.

The Singapore secondary school syllabuses define mathematical modelling as follows:

Mathematical modelling is the process of formulating and improving a mathematical model to represent and solve real-world problems.

Through mathematical modelling, students learn to use a variety of representations of data, and to select and apply appropriate mathematical methods and tools in solving real-world problems. The opportunity to deal with empirical data and use mathematical tools for data analysis should be part of the learning at all levels. (Ministry of Education, 2006, p. 4)

It is helpful to think of this process diagrammatically. A typical representation of the thinking involved is shown in Figure 1, taken from the seminal Australian document, *A national statement on mathematics for Australian schools* (Australian Education Council, 1990, p. 61).

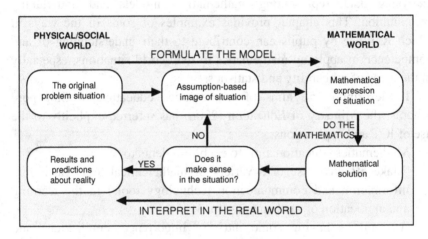

Figure 1. Mathematical modelling according to *A national statement on mathematics for Australian schools*

It is important to note that the use of ICT in such a formulation occurs only on the right of the diagram: representing a model within an ICT context and using the ICT to "do the mathematics" are the key features. The rest of the process of mathematical modelling involves thinking skills. The diagram also makes it clear that modelling involves encountering and using real data of some kind.

2 A Variety of ICT Tools

There is a variety of ICT tools that pupils might find helpful for applications of mathematics. An object is only likely to be helpful as a tool, however, if attention is given to learning how to use it effectively, efficiently and autonomously, so that attention is needed for this in the classroom. Indeed, it seems important for attention (and time) to be given to the development of expertise with appropriate ICTs as part of the tools of the mathematical trade for all pupils. Brief comments about the range of possible tools is offered below.

2.1 *Calculator*

A standard calculator (sometimes – inaccurately – referred to as a four-function calculator) will allow pupils to undertake numerical computations when the tasks are beyond their mental capabilities, or when accurate results (rather than rough approximations) are needed. While in many cases, approximate results are sufficient at first, it will rarely be the case that such a calculator will be sufficient.

2.2 *Scientific calculator*

A scientific calculator usually allows for 'table' functions (such as trigonometric or logarithmic functions) to be readily evaluated, obviating the need for a book of mathematical tables. It will also extend the range of numbers that can be accommodated, to very large and very small numbers, because of the availability of scientific notation. In addition, they permit some more sophisticated mathematical operations such as raising numbers to powers and finding roots; recent models also allow for some elementary data analysis, usually restricted to finding sample statistics such as means and standard deviations. For these reasons, such calculators have been in widespread use in school curricula in Australia now for some thirty years or so. However, scientific calculators essentially do not provide tools that are much more sophisticated than elementary calculators and offer little more to applications and modelling than tools for calculation of numerical results.

2.3 Graphics calculator

In sharp contrast to scientific calculators, graphic calculators offer pupils significant opportunities to engage in mathematical modelling, as they permit pupils to examine information in a range of ways (Kissane, 2007) and, critically, *experiment* with mathematical objects. Thus, for example, not only can pupils evaluate a mathematical function at a point (as they can on a scientific calculator), but they can do so at a set of points to generate a table of values. Similarly, rather than being restricted to numerical values, pupils can represent a function graphically (of considerable importance in trying to choose a mathematical model for data). In addition, it is relatively easy for pupils to change a function and see the consequences for its numerical values or its graph. Collectively, these multiple representations of mathematical functions are often described as "the rule of three", referring to the symbolic, numerical and graphical representations. Clearly, such capabilities extend the range of pupil activities for modelling very considerably.

Newcomers to such technology frequently think that the mathematical capabilities of graphics calculators are mostly concerned with this idea of representing functions, especially drawing graphs of functions; indeed, some refer to graphics calculators as 'graphing' calculators, sympathetic with this view. However, the technology of graphics calculators is much more powerful than this restricted view suggests, and offers pupils significant experimental possibilities in many domains other than graphing of functions. As a reflection of this, only about a quarter of Kissane and Kemp (2006) is concerned with function graphing. Thus graphics calculators offer significant capabilities in equation solving, sequences and series, elementary calculus, geometry and probability, independent of the idea of function graphing. For this reason, it is arguable that the term 'calculator' captures only poorly the range of capabilities, although it is difficult to imagine a new term gaining currency.

Most importantly for the present purpose, graphics calculators offer significant new opportunities for pupils to engage in data analysis, a key element of mathematical modelling with real data. Where a scientific calculator merely provides some numerical statistics for a data set, a graphics calculator allows for models for the data to be constructed, modified, examined and tested, not unlike the capabilities offered by a statistics package.

2.4 *CAS calculator*

In recent years in some countries (including Australia and the USA), graphics calculators with computer algebra system (CAS) capabilities have increasingly become available. These allow for exact mathematical processes such as manipulating algebraic expressions (e.g., factoring and expanding), solving equations, differentiating functions, evaluating integrals and solving differential equations to be handled by the calculator. In several Australian states, such tools are now regarded as standard equipment that senior secondary school pupils are expected to use well and are permitted to use in important examinations, such as those at the end of secondary school.

As far as modelling is concerned, access to such tools allows for experimentation with symbolic representations of models to be undertaken. This allows pupils to model situations in general rather than in particular, provided they have a sufficient understanding of the mathematical ideas involved.

Recent hand-held devices include a very wide range of capabilities, including CAS capabilities, making them self-contained mathematical tools of great sophistication. The best known examples at present are Casio's *Classpad* and Texas Instruments' *TI-Nspire*, both of which are popular in schools in Australia. Interestingly, neither of these is usually referred to as a 'calculator' by their respective manufacturers, reflecting the very many tools beyond mere 'calculation' that they offer to pupils. Such tools can gain currency in schools only when the school curriculum has adjusted to them, however, so that it seems prudent to restrict attention in this chapter to hand-held technologies that do not incorporate CAS.

2.5 Computer software

The major advantage that calculators hold over computers is that they are portable. Being small, light and battery-powered means that they can be easily taken from class to class, between home and school, from classroom to examination room. These are significant advantages for the design of a curriculum.

Computers tend to be larger, but with significant strengths not available to calculators. These include much larger memories, colour displays, appropriate software and telecommunications capabilities. These capabilities can be very helpful for mathematical modelling, and it seems highly unlikely these days that professionals engaged in applying mathematics to real world problems would do so without significant ICT use.

Generic software such as Microsoft *Excel*, has the advantage of being widely available (as it is often bundled with computer purchases) and also widely used outside schools (in businesses, for example). Although the software was not designed for mostly mathematical purposes, and certainly was not originally intended for educational uses, it has often been used creatively by teachers and even by statisticians to handle tasks involving modelling. A spreadsheet is especially useful when very large data sets are involved (although this is rarely the case for secondary school), as it has much better data handling capabilities than a calculator (although somewhat inferior to software designed for statistics).

In recent years, some mathematical software has been developed explicitly for educational use. Outstanding commercial examples for statistics include *Tinkerplots* and *Fathom*, while outstanding examples for geometry include *Geometer's Sketchpad*, *Cabri Geometry* and *Cabri-3D*. Because these have been designed for educational use, they also have associated materials of various kinds available for teachers, which is an important practical consideration. More recently, free software such as *GeoGebra* has been developed, and considerable work has been done internationally to share educational uses of this software with a wide audience of teachers.

When schools have the necessary facilities, there is much to be gained by providing pupils with access to software packages of these kinds, although it seems unwise to rely entirely on such software to support pupil work, at least until such time as computers themselves become as ubiquitous, inexpensive and acceptable to examination authorities as calculators are.

For pupils engaged in applications of mathematics, computer presentation software (such as Microsoft *PowerPoint*, typically included as part of a computer bundle of software and thus widely available) is also important, as it allows them to describe their work to fellow pupils or to others. If mathematical modelling and applications are to move beyond 'exercises' for pupils and become a more prominent part of pupils' work (for example, in the form of projects or activities), software of this kind will allow the pupils to consider the key elements of communication and justification of their findings to an interested audience.

2.6 *The Internet*

In recent years, the Internet has become a prominent part of people's lives in many countries, certainly including Singapore as well as Australia. Consequently, there are many ways in which the Internet might be used for educational purposes, with a typology of these offered by Kissane (2009); for this reason, the author maintains structured links to suitable websites for mathematics pupils and their teachers at the address above. As far as applications and modelling are concerned, as well as providing a range of current examples and some useful tools, the Internet may be used in schools to provide a source of authentic data that pertain to questions of interest to pupils.

There is not space in this chapter to exemplify all of these ICT tools for applications of mathematics. Instead, we focus attention on the important areas of statistics and probability. Partly, this recognises the primacy of these aspects of mathematics in the Singapore secondary school syllabus descriptions and partly it is because the use of ICT has transformed this area of mathematics in recent years.

3 Modelling with a Calculator

As their name suggests, calculators are helpful for doing calculations. Many of the everyday applications of mathematics referred to in the secondary school syllabuses (Ministry of Education, 2006) require relatively little computations and there is much to be said for pupils learning to apply their mathematics approximately and mentally to the everyday world. However, when the numerical data go beyond what can be comfortably handled mentally, when many calculations are needed, or when more precision is appropriate than good approximations can provide, use of ICT for computational help seems unavoidable. The alternative is to restrict applications of mathematics to invented data and further increase a common pupil view that mathematics is not really helpful for understanding the world.

Consider the use of mathematical ideas to model and to understand the changing population of Singapore. Statistics Singapore (2009) has provided relevant data, accessible to pupils. These data can be entered into a scientific calculator and a suitable line of good fit obtained, but the exercise of doing so is unlikely to add much understanding of what is happening. On a graphics calculator, data are stored, and thus can be manipulated in some ways. For example, Figure 2 shows some data and a least squares regression line of best fit for the total population from 1980.

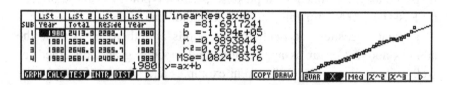

Figure 2. Using a graphics calculator for analysing Singapore's population

The analysis suggests that the data are reasonably well-approximated by a linear function, as reflected in the high value of r^2 and the visual appearance of the scatterplot and the least squares line.

The quality of the modelling in this case can be explored by imagining that the analysis was conducted from 1980 to 1990 and then using the results to predict the population at a later time (such as 2008, the most recently available data). Figure 3 shows the calculator being used in this flexible way, tabulating a linear function to arrive at a predicted population of 3.89 million people in 2008.

In fact, the population of Singapore in 2008 was substantially (almost a million) more than this prediction, suggesting some caution about using a simplistic linear model to describe something as complex as population growth, and helping pupils to realise some of the realities

 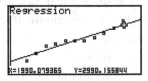

Figure 3. Predicting Singapore's population in 2008, based on data from 1980-1990

of dealing with real data. Once the data are stored in a graphics calculator, pupils can explore them in various ways, helping to understand more fully the real world they represent. In this case, further explorations might involve considering subsets of the data or involve applying more sophisticated models than the simple linear regression, consistent with the idea of successive improvements to a mathematical model reflected in Figure 1 above.

4 Simulation and Modelling

While many practical applications of mathematics rely on accessing authentic data, there are others for which the simulation of random data is more appropriate. A very useful facility on all mathematical ICT (including in particular scientific and graphics calculators as well as statistical and spreadsheet software) involves the capacity for random generation of data. Typically, this is based on psuedo-random numbers, uniformly distributed on the interval (0,1).

A good example of this, readily accessible to school pupils via simulation, involves the birthday problem. This interesting problem concerns the probability that a randomly chosen people group of a certain number of people will include at least two people with the same birthday. Determining this probability is both practically difficult (because of the large number of calculations) and conceptually difficult (as it requires a good understanding of independence and laws of probability). To generate data that represent birthdays, a transformation of random numbers from (0,1) to random integers on [1,365] is needed, ignoring for the sake of simplicity the unusual birthday of 29 February.

Although some devices handle such a transformation automatically, the elementary mathematical idea involved is too important to omit from any modern school curriculum. If RAN# represents a number in (0,1), then the transformation:

Int (365*RAN# +1)

produces a random integer on [1,365], where Int refers to the integer part of a number. Generation of a set of 30 'birthdays', each of which is the number of a particular day of the year, is relatively easy on a graphics calculator, as shown in Figure 4.

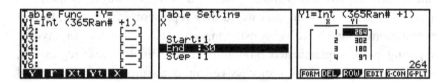

Figure 4. Generation of 30 birthdays

Once the data are generated, a sort of the birthdays facilitates a quick check for any matches. In the example shown in Figure 5, there is a match, indicating that two of the thirty simulated people in this set have the same birthday.

Figure 5. Sorting data to find a match

This example illustrates some important points: applications of mathematics using ICT do not always use the same mathematical approaches as more formal treatments; relatively unsophisticated mathematics can be used to address an everyday question; understanding how to use an ICT tool allows it to be used flexibly for modelling.

Simulation practices can be usefully extended to model more sophisticated situations. To illustrate, consider a Plane Bookings problem, of determining the most appropriate number of bookings an airline ought take for a small commuter plane, recognising that it is inefficient to fly the plane without all seats being occupied, but that some people make bookings and fail to turn up for the flight. Problems of this kind can be simulated, after making suitable assumptions, and again using an elementary transformation of random data. To be concrete, suppose that on average one out of six passengers is a 'no-show' (a figure not markedly different from some actual data) and that the plane has only 20 seats.

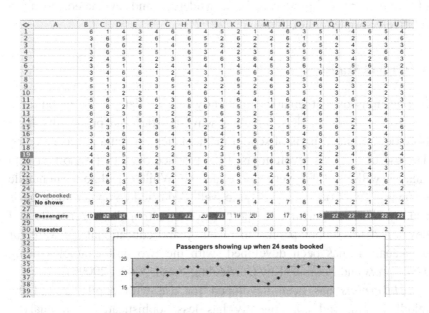

Figure 6. Simulating flight bookings on a 20-seat plane

Simulation can be used to explore the consequences of various possible strategies for this situation, which is quite difficult for school pupils to tackle using school mathematics. Figure 6 shows an *Excel* spreadsheet exploring successive examples of booking 24 seats, to determine the number of seats occupied, and how often the plane is overbooked (so that there are unseated passengers). The mathematics behind this spreadsheet comprises the simple transformation of Ran# to Int (6Ran# + 1), which generates dice rolls. For the purposes of the application, the spreadsheet regards a result of 5 as representing a no-show.

Simulations of this kind allow pupils to see that, while theoretically booking 24 seats will on average result in 20 passengers taking their seats, there will be times when the planes are overbooked and also times when the plane is still not filled. What the simulation provides that a formal analysis of probabilities will not, is a sense of how often various phenomena occur, so that a realistic appraisal of the proposal can be considered. The consequences (both in the real world and in the model) of various possibilities can be considered, and evaluated, using a relatively simple tool of this kind.

5 ICT and Statistics

Applications of mathematics frequently rely on data, both for building and for verifying models, which is why statistics and probability have become prominent parts of school mathematics curricula in recent years. The use of ICT has revolutionised the practice of statistics, so that data analysis without ICT is very hard to defend. While calculators can handle some aspects of data analysis quite well, dedicated computer software is usually more powerful and more flexible, although is generally not developed for school pupils. In recent years, educational software for data analysis has been developed, with the two best examples being *Tinkerplots* and *Fathom* (Key Curriculum Press, 2009a; 2009b).

Tinkerplots has been developed for use with pupils around the middle years, and consequently has less sophisticated data analytic procedures inherent in it than does *Fathom*. However, it provides pupils with powerful ways of engaging in data analysis, with the aim of understanding the data well enough to 'tell the story' associated with them. This is especially important for analysing data obtained by pupils themselves to answer questions of their own interest.

An excellent example of this concerns pupil backpacks, with data collected from a school and made available for illustrative analysis with the software. (Readers can download evaluation versions of the software, including the associated datafiles, from the web link provided by Key Curriculum Press (2009a).) Data comprise information obtained from 79 pupils of various classes, including their weight and the weight of their backpack, and were obtained from pupils in a US school, using a set of scales. Of key interest is the comparison between pupil weight and backpack weight. The role of the software is to facilitate data analysis, so that pupils are not distracted by the tasks of calculating statistics or drawing graphs, but instead can focus on the meanings behind the data. Each case provides information on various pupil attributes such as class, body weight, backpack weight, and so on. Sorting the cases is a primal act of analysis, and the software provides very intuitive ways for pupils to do this to reveal important relationships.

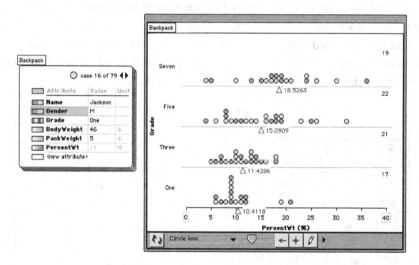

Figure 7. Exploring *Backpacks* (Key Curriculum Press, 2009a)

The use of software of this kind allows for less emphasis on sophisticated quantitative analyses and more emphasis on the *story* of the data. The story can be constructed by choosing suitable ways of representing the data visually and numerically. In this case, Figure 7 shows the extent to which the weights of backpacks, as a proportion of the weights of the pupils, seem to grow substantially as pupils progress through school, with an alarming number exceeding the recommendations of health authorities (of a maximum of 15%). The diagram shows mean percentages and also allows pupils to explore other questions of interest, such as differences between boys and girls.

While *Tinkerplots* is described by its publishers as suitable for grades 4-8, it arguably extends beyond both extremities. *Fathom* (an earlier statistics package from the same developer) is also designed for educational purposes, but with a slightly older focus, including inferential statistics and formal mathematical procedures such as linear regression.

Although some of these procedures are accessible via graphics calculators, as illustrated earlier, computer software provides more analytic opportunities and better visual representations. To illustrate, Figure 8 shows representations of the Singapore population data, including the least squares regression line shown previously. Notice how well the 'least squares' concept is demonstrated in the graph: where a calculator provided only the numerical results, the software also allows the underlying concept to be seen. The 'line of best fit' is the line that minimises the sums of the areas of the squares shown.

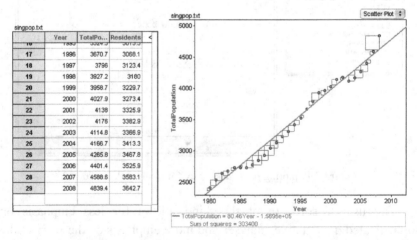

Figure 8. Analysing Singapore population data using *Fathom*

Using ICT in Applications of Secondary School Mathematics   193

In addition to improving the understanding of key ideas, a tool like *Fathom* allows pupils to more easily explore underlying patterns in the data. In this case, graphs of the two population variables: the 'Residential" population and the 'Total' population illustrate that the former is much more linear than the latter, as shown in Figure 9. That is, it seems as if Singapore's residential population can be predicted fairly well with a linear model perhaps because it depends mostly on a relatively stable rate of natural phenomena such as birth and death rates. In contrast, the total population is harder to predict successfully, as it presumably involves other factors such as migration, government policy decisions, economic influences, and so on. This accounts for the rather poor prediction noted with the graphics calculator model earlier. Although similar analyses can be conducted on the less sophisticated calculator, the environment provided by software like *Fathom* requires less effort on behalf of the user to engage in exploratory modelling of these kinds.

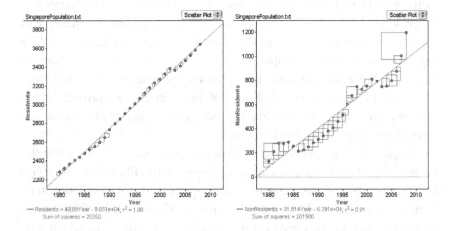

Figure 9. Analysing total and residential Singapore population using *Fathom*

Data analysis is most powerful when pupils are addressing problems that matter to them. They can design and undertake their own data collection, then use software such as *Tinkerplots* or *Fathom* to store and analyse their own data. Use of the software does not require high-level

ICT skills, and pupils will incidentally learn important lessons associated with statistics such as data editing, dealing with missing cases, identifying and correcting entry errors, and so on.

6 Using the Internet

The Internet has become one of the most important aspects of ICT in recent years, and also has some connections with mathematical modelling and applications. Kissane (2009) explores some of the ways in which the Internet is significant for both pupils and teachers of mathematics, several of which are of particular importance for applications of mathematics and mathematical modelling. The author's website systematically catalogues some of the ways in which pupils might learn about mathematics through use of the Internet. For example, one category of use involves 'Reading interesting materials', in which there are websites identified that describe current examples of mathematical modelling and the use of mathematics in applications of many kinds. The *Plus* magazine from the UK is one of these, with regular contributions from people using mathematics, ideally suited for sophisticated secondary pupils. Similarly, the Australian Academy of Science's *Nova* site contains many examples of mathematical modelling, described in some detail. Websites of these kinds can provide recent material for pupils, reinforcing the perspective that mathematics is a valuable tool to address modern problems.

Applications of mathematics involving simulations are best handled by the use of ICT, as the earlier examples have illustrated. In recent years, a number of excellent tools and applications of simulation have been made available through the Internet. In Kissane's (2009) typology, some of these take the form of 'Interactive opportunities', for which links are provided. For example, the outstanding National Library of Virtual Manipulatives provides several examples in Data Analysis and Probability for pupils to use for simulation, allowing data to be generated, organised and analysed efficiently and thus helping pupils to see how randomness can be understood and used for practical purposes. Similarly, a search of the NCTM *Illuminations* site in the Data Analysis and Probability area provides a rich set of resources for both pupils and their teachers.

As a third example, the remarkable *Nrich* site also has many simulation tools among its regular resources for pupils and teachers. An important new kind of resource offered by *Nrich* are teacher and pupil packages, which can be downloaded for offline use, particularly helpful when access to the Internet is not readily available in a classroom. For example, one of these (Nrich, 2009) concerns probability, and offers a collection of several Flash applets that can be used by the whole class (e.g. on an interactive whiteboard or a whole class display) or used by small groups or individual pupils. Figure 10 shows one of these applets, allowing for efficient simulation of tossing a pair of discs, differently coloured on each side, and recording successive results in a powerful way (using the sheet developed by the DIME project). In this case, the use of ICT has made it easier to use good ideas for pupils to explore simulation in a way that was developed many years ago, but which was difficult to implement for practical reasons.

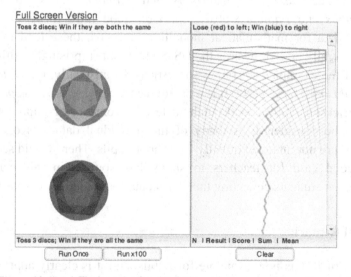

Figure 10. Nrich Flash applet for simulating coin tosses and recording results

As well as providing pupils with good examples and with tools for simulation and modelling, the Internet can also be used as a source of data. Most data on the Internet are not case-based data, but are summary data, which often provide less opportunity for pupil exploration. An example of case-based data is the Census At School project, administered by the Australian Bureau of Statistics (2009) and which is intended for pupils themselves to use. After a large group of pupils has completed census-like surveys, a large body of data has been stored on the Internet. Pupils are able to download a random sample of these data for analysis. Significantly, since each random sample is different, not only do pupils thus get an opportunity to explore relationships among real data, but they also have an opportunity to see how different their conclusions are from those of other pupils, and in the process begin to learn about the pitfalls (as well as the benefits) of random sampling. Although there is not yet a Singaporean version of this project (and consequently the data may be less directly relevant culturally to Singapore pupils), much is to be gained by pupils using this kind of data set to understand difference between themselves and others.

Some data are available on the Internet for direct downloading into software packages like *Tinkerplots* and *Fathom*. Indeed, both the *Tinkerplots* and *Fathom* websites provide good links to various data sources data for mathematical modelling, although these tend to be a little culturally biased towards the US context, not surprisingly. Although some of the examples are rather sophisticated statistically, the *Data and Story Library* (DASL) (2009) is a useful data source, as is *The eeps Data Zoo* (Erickson, 2007). With such materials, contexts of pupil interest need to be considered, as many of the individual data sets on these websites are not intended directly for school pupils. There would seem to be a good case for teachers to share data sources in this way, to supplement students collecting their own data for modelling purposes.

7 Final Remarks

A range of ICT is now available for pupils, and it is clearly appropriate for the mathematics curriculum to take advantage of this, as the Ministry of Education (2006) has noted. The use of ICT for applications and modelling is constrained by facilities, the curriculum and the time available. Some examples of ICT (such as calculators) are likely to be available everywhere, and to meet many needs. These are most likely to be accommodated by external examination constraints, an important consideration for classroom use. Other manifestations of ICT, such as specialist software, are more powerful, but usually require more resources to use effectively.

This chapter has suggested that ICT can play a substantial role in improving the statistics and probability components of the school curriculum in particular. In ways that were not previously possible, pupils can engage in genuine data analysis, and explore questions of interest to themselves. They can also explore chance phenomena through the use of simulation, not previously accessible without access to ICT.

While teachers do not need extensive backgrounds in the use of ICT to take advantage of these opportunities, they do need support, professional development and time to learn how to use the ICT effectively and efficiently. As calculators and software become more user-friendly, and as the Internet continues to offer many new materials for teachers and their pupils, teachers can expect that the opportunities for sound use of ICT for mathematical modelling and applications will increase.

References

Australian Bureau of Statistics. (2009). *Census at School*. Retrieved November 8, 2009, from http://www.abs.gov.au/websitedbs/cashome.NSF/home/home
Australian Education Council. (1990). *A national statement on mathematics for Australian schools*. Australia, Melbourne: Curriculum Corporation of Australia.
Erickson, T. (2007). *The eeps Data Zoo*. Retrieved November 8, 2009, from http://www.eeps.com/zoo/index.html
Key Curriculum Press. (2009a). *Tinkerplots® Dynamic data Exploration*. Retrieved November 8, 2009, from http://www.keypress.com/x5715.xml

Key Curriculum Press. (2009b). *Fathom™ Dynamic Data Software*. Retrieved November 8, 2009, from http://www.keypress.com/x5656.xml

Kissane, B. (2007). Hand-held technology in secondary mathematics education. In S. Li, J. Zhang & D. Wang (Eds.) *Symbolic Computation and Education* (pp. 31-59). Singapore: World Scientific.

Kissane, B. (2009). What does the Internet offer for students? In C. Hurst, M. Kemp, B. Kissane, L. Sparrow & T. Spencer (Eds.) *Mathematics: It's Mine: Proceedings of the 22nd Biennial Conference of the Australian Association of Mathematics Teachers* (pp. 135-144). Australia, Adelaide: Australian Association of Mathematics Teachers.

Kissane, B. & Kemp, M. (2006). *Mathematics with a graphics calculator: Casio fx-9860G*. Australia, Perth: Mathematical Association of Western Australia.

Ministry of Education. (2006). *Mathematics syllabus – Secondary*. Singapore: Author.

Nrich. (2009). *Experimenting with probability*. Retrieved November 8, 2009, from http://nrich.maths.org/public/viewer.php?obj_id=5541

Statistics Singapore. (2009). *Time series on population*. Retrieved November 8, 2009, from http://www.singstat.gov.sg/stats/themes/people/hist/popn.html

The Data and Story Library. (2009). Retrieved November 8, 2009, from http://lib.stat.cmu.edu/DASL/

Chapter 11

Developing Pupils' Analysis and Interpretation of Graphs and Tables Using a *Five Step Framework*

Marian KEMP

In the last decade there has been a substantial increase in the exposure of most people across the world to quantitative information in both print and digital media. This information is derived from a variety of sources including surveys, opinion polls, scientific experiments and research in the health sciences and it is presented as numbers in text, or in graphs, charts or tables. Analysing and interpreting statistical information in these forms can be quite complex, and are even more difficult when the presentations are of poor quality and potentially misleading. To fully understand the information conveyed in text, tables and graphs requires transformations between the real world and the mathematical world, and back again. It requires an application of mathematical and statistical concepts and skills to the contextual material. This chapter will model the use of a *Five Step Framework* designed to assist pupils to develop those skills and strategies required to interpret data in tables and graphs encountered across curriculum areas, including National Education in school.

1 Introduction

Changes in technology over the last twenty years have led to increases in the speed and extent of communication of information across the world.

Even though a presentation in the form of text, graphs or tables may look quite simple the analysis and interpretation of much of this information can be quite complex. As Ben-Zvi and Garfield (2004, p. 3) point out:

> Quantitative information is everywhere, and statistics are increasingly presented as a way to add credibility to advertisements, arguments or advice. Being able to properly evaluate evidence (data) and claims based on data is an important skill that all students should learn as part of their educational programs.

Indeed, people need to be aware that data is collected from a variety of sources, some more credible than others, and that authors may use selected data to support their own points of view, often without disclosure. For example, Best (2001, p.18) emphasises that:

> ... we need to understand that people debating social problems choose statistics selectively and present them to support their point of view. Gun-control advocates will be more likely to report the number of children killed by guns, while opponents of gun-control will prefer to count citizens who use guns to defend themselves from attack.

Thus, people need to think carefully about the arguments presented in the media and the evidence that is given to support them. They also need to be aware that there are often defects in the graphs and tables that can unintentionally, or intentionally, mislead or misinform the reader. In this area Gal (2003) proposes the use of press releases in the media to help develop awareness and skills. This chapter proposes that all pupils should gain sufficient understanding and skills to analyse and interpret data that they encounter, whether it is in well, or poorly constructed graphs or tables.

This chapter will focus on the use of a *Five Step Framework* (Kemp, 2005) that was based on the SOLO Taxonomy (Biggs & Collis, 1982) and designed to enable teachers and students in the mathematics classroom to develop strategies for interpreting data in graphs and tables. It will explore some examples of data representation chosen as suitable for study in the context of the National Education curriculum.

2 Data Analysis in the Mathematics Curriculum

The Ministry of Education (2006a) in Singapore acknowledges the importance of data analysis by including it in *Statistics and Probability* sections in the 2007 secondary mathematics syllabuses at all levels. In addition, in the section of the syllabus documents called *Applications and Modelling* they emphasise the importance of the application of mathematics to "real-world problems" and expect that "students learn to use a variety of representations of data" (p. 4) at all levels. Notably, the Normal (Technical) Level Mathematics includes *Integrative Contexts* from Secondary Two to Four that include reference to everyday statistics and interpretation of data in tables and graphs.

In Singapore it is expected that all teachers be involved in National Education. For example, National Education has been explicitly incorporated into the Social Studies Syllabus (Lower Secondary Normal (Technical)) in conjunction with the need to develop the ability to critically interpret data by including the following in the general aims:
- To develop skills of information gathering, data analysis and evaluation which are necessary for learners of the 21st century
- To inculcate in students a sense of appreciation and responsibility for the society and environment
- To develop students into informed citizens who will be able to have a better understanding of national and world issues (Ministry of Education, 2006b, p. 1)

Wong (2003) gives some useful examples that mathematics teachers can use in their classrooms for developing National Education from a range of contexts. In addition, mathematics teachers can play a substantial part in this development by using authentic data from trusted websites in the mathematics classroom and using the opportunities offered to discuss issues relevant to National Education. It follows that mathematics teachers should play a strong part in the development of data analysis and evaluation in this context.

Data analysis involves a range of skills. One of these is analysis and interpretation of graphs that is often considered to be easier than interpreting tables of data (Wainer, 1992). However, Wu and Wong

(2009) have analysed a number of studies indicating that although pupils can read simple values from statistical graphs, they have difficulties in developing valid interpretations of situations from the graphs. This highlights the need to find a way to help pupils develop a range of data analysis and interpretation skills.

3 A Five Step Framework

The *Five Step Framework* (Kemp, 2005) summarised in Table 1, is based on the SOLO taxonomy developed by Biggs and Collis (1982). The taxonomy is itself based on their research that identified a typical developmental transition from simple to complex responses to a situation. The *Framework* reflects this hierarchy in the way the Steps are formulated and provides a generic template for teachers to assist their pupils to develop strategies for interpreting data in graphical or tabular form. The *Framework* can be applied to both simple and complex graphs and tables. The level of complexity and the context of the graph or table need to be selected according to the types of materials being studied by the students and their mathematical backgrounds. The *Framework* highlights the progressive kinds of questions pupils need to ask themselves when analysing and interpreting graphs and tables.

Research has shown that the use of the *Framework* in workshops in a university in Western Australia improves the abilities of undergraduate students to interpret tables and graphs (Kemp, 2005; Kemp & Bradley, 2006). The *Framework* has also been used over several years in workshops with primary, secondary and tertiary teachers. These workshops have been located in Australia, Singapore, Thailand, the Philippines, the Czech Republic, Germany and Mexico. Informal feedback has confirmed that there is a need for work of this kind in classrooms, and that teachers who have used the *Framework* have reported that it acts as a useful scaffold. The scaffolding takes the form of questions appropriate to the steps and to the context. Examples of these are included in this chapter. For younger or less experienced pupils teachers might scaffold them to use only Steps 1, 2 and 3. Once the pupils have developed some understanding of the strategies within the

Framework and some skill in using the steps they can develop their own questions to interpret the tables and graphs that they encounter.

Table 1

Five Step Framework for analysing and interpreting graphs and tables

Step 1: Getting started
Look at the title, axes, headings, legend, footnotes and source to find out the context and expected reliability of the data. This involves seeking information about sampling methods, sample sizes and predicted accuracy. For surveys and opinion polls this also involves looking at the questions asked. Look up any words that are new to you.
Step 2: WHAT do the numbers mean?
Make sure you know what all the numbers (percentages, '000s, etc) represent. Look for the largest and smallest values in one or more categories or years to get an idea of the range of the data.
Step 3: HOW do they differ?
Look at the differences in the values of the data in a single data set, row, column or part of a graph. Repeat this for other data sets. This may involve changes over time, or comparisons within categories, such as male and female, at any given time.
Step 4: WHERE are the differences?
What are the relationships in the table or graph? Use your findings from Step 3 to help you make comparisons between columns or rows in a table, or parts of a graph, to look for similarities and differences.
Step 5: WHY are there differences?
Look for possible reasons for the relationships in the data you have found by considering cultural, societal, environmental, political and economic factors. Think about sudden or unexpected changes in terms of state, national and international policies or major events.

4 Analysing and Interpreting Graphs and Tables

The Singapore Mathematics syllabuses require that students learn to construct tables and graphs as well as interpret them (Ministry of Education, 2006). The development of the practical skills needed in the construction of tables and graphs is beyond the scope of this chapter. However, the author proposes that through critical analysis of tables and graphs through the use of the *Framework* pupils can develop some understandings of the purposes of different kinds of graphs and tables and so make more informed and sensible choices of the kinds of tables or graphs when they present their own data.

4.1 *Statistical graphs*

The benefits of using graphical displays of information have been acknowledged in the literature for many years (Tukey, 1977; Wainer, 1992). There are examples in the literature investigating the ways in which graphical displays can be misleading, with or without intent, (e.g. Dewdney, 1993; Watson, 2006). Although some people assert that people can read graphs and detect the lies (Wainer, 1992; Tufte, 1983) Watson (2006) stresses that graphs and tables in the media can be quite misleading and states that in the school curriculum "there is a conscious effort that needs to be expended to achieve skills that will provide the lie detectors required by statistically literate adults" (p. 55).

In Australia (Watson, 2006) and in Singapore (Wu & Wong, 2009) the mathematics curricula acknowledge the importance of developing students' abilities to construct, understand and interpret graphical displays. Wu and Wong (2009) cite studies which found that, although students can read simple values from statistical graphs, they have difficulties in developing valid interpretations of situations from the graphs.

Students need to be provided with a wide range of learning experiences with constructing and interpreting graphs in a number of contexts to gain some experience and knowledge about the way that appropriate graphs are selected. For example, line graphs for trends, pie graphs for proportions of the whole, dot plots for comparing in small groups, column or bar graphs to compare discrete values, histograms for continuous data and scatter plots for bivariate data. As Watson (2006, p. 91) points out:

> Working among various graphical forms with students provides links among skills and mathematical concepts related to graphing. It also helps to develop appreciation for which types of graph assist in displaying the message to be told.

Experience with graphs from reputable sources interrogated using the *Framework* should help in this regard.

4.2 Tables of two kinds

There are two main kinds of tables that people encounter, both of which require skills to make sense of them. Firstly, there is the kind of table that is consulted to gain specific information. This is the kind of table referred to by Mosenthal and Kirsch (1998, p.639) as 'reading-to-do'. These include tables like those with instructions for dosages of medicine depending on age or weight and those on food and health supplement packaging as well as a range of transport timetables. Although there can be some difficulties in reading these tables and people need to learn how to do so there is usually some experience of this traditionally included in the secondary school curriculum (Kemp & Kissane, 1990).

Secondly, there is the kind of table that includes collected, produced and collated data that can be used to develop an understanding of the associated context. This is the kind of table referred to by Mosenthal and Kirsch (1998, p.639) as 'reading-to-comprehend'. Examples of these tables include data on unemployment, traffic statistics, poverty, disease, and educational achievement that would be found in newspapers, journals and reports. Many people skim over these kinds of tables, relying on the possibly biased interpretations of the authors of the accompanying text (Chapman, Kemp & Kissane, 1990).

Tables are often hard to understand because they have not been well constructed (Ehrenberg, 1981). However, it is important that pupils develop the approriate strategies and skills for interpreting tables that are poorly constructed as well as those that are well-constructed. This is because there are reasons for using a table to communicate data (Koschat, 2005). Firstly a table presents data, or a summary of that data, in their original form. Secondly, people can easily use the data and convert to other forms such as a graph or a model if they wish to do so, but this is not easy in reverse. Thirdly, it can quite often be the case that the reader wants to see and interpret the actual numbers. If people are to be able to make use of tables for any of these reasons the tables should be constructed to facilitate this. The reconstruction of a table can make a big difference to the readability and understanding of the table. Wainer (1992) suggests that to make tables readable and logical the rows and columns should be put in an order that makes sense, making use of marginal averages or totals; the numbers to be compared should be put in columns with the highest at the top; rounding should be done to two effective digits and the layout should facilitate comparisons.

5 Using the *Five Step Framework* in the Singapore Context

In this section examples of activities are provided to indicate how the *Five Step Framework* can be used in the mathematics classroom to help pupils to develop strategies for analysing and interpreting data. Given the importance that is placed on National Education in Singapore (Ministry of Education, 2009), the activities using the *Framework* have all been chosen to provide contexts for pupils to think about Singapore and other countries in relation to Singapore. Naturally, there are many other contexts in which the *Framework* can be used. The activities include the analysis of line graphs, column (bar) graphs, a simple table and a more complex table.

As indicated in Table 1 the *Framework* outlines a series of hierarchical steps through which the analysis of graphs and tables naturally proceeds. Pupils can be assisted to understand and use this hierarchy by providing them with a series of specially designed focus questions or tasks. However, it is productive to initially ask the pupils to interpret data in different forms without the aid of the *Framework* to help them to realise that high quality interpretation of data is actually quite difficult and that it involves strategies, skills and persistence. Therefore it is suggested that pupils are asked to look at the graphs or tables before being offered the scaffolding of the *Framework*.

When using the *Framework* initially to develop the skills and strategies the teacher would design appropriate questions or tasks for each Step. Step 1 type questions lead the pupil to identify the context and main features of a graph or table. Step 2 type questions are used to ensure that pupils understand the kinds of numbers being used, (e.g., units, '000, %s) and that they can answer questions about single points in a table or on a graph. Step 3 type questions are to prompt comparisons in a single context, one row or column in a table or one category in a graph. Step 4 questions require comparisons of a more complex nature involving two or more columns or rows in a table or two or more categories in a graph. Step 5 requires pupils to consider the data in the light of cultural, societal, environmental, political or economic factors.

The number of steps to be attempted by pupils and the complexity of the questions or tasks will initially be determined by the teacher based on the mathematical backgrounds and sophistication of the pupils. Once the pupils become adept at using the steps it is expected that they will determine their own questions depending on their requirements and the level of complexity of the graph or table. In this chapter typical questions for the graphs and tables are provided. Teachers should note that these are provided as a guide and that their own use of the *Framework* will vary in different classrooms.

Activities One, Two and Three involve analysis of data collected as part of the *National Youth Survey* 2005 and presented in *Social Participation and Active Citizenship* (Chong & Chia, 2006). The *National Youth Survey* 2005 collected and analysed data on a wide range of issues. The data collection was done with the following research methodology.

- A stratified probability sampling method was used to obtain the sample for the National Youth Survey. The Master sampling frame of the Department of Statistics was used. Polling Districts were first stratified into three housing types: public housing, private housing and others.
- The first stage of the stratification involved the systematic probability selection of Polling Districts proportional to the sizes of the Polling Districts. The second stage was the selection of household units within each polling district using a random start. The key object of the chosen sampling methodology was to obtain a representative sample of young Singaporeans. The sampling frame covers Singaporeans and Permanent Residents aged 15 to 29 years (p. 16).

However, it is not always the case that samples are randomly selected (in this case stratified random sampling was used). Given that for any valid statistical conclusions to be drawn the sample must be random (Watson, 2004), it is essential that in the classroom students develop an understanding of the concept of a random sample and the various ways that collection of a simple random sample can be achieved. Discussion in the classroom could focus around the question of sample size and the disadvantages and advantages of small and large samples (Chance, delMas & Garfield, 2004). Activity Four uses data from the *Yearbook of Statistics Singapore* (2009) concerned with *Crime Cases* in Singapore recorded in different ways.

5.1 *Activity one*

Analyse and interpret the data on Singapore youth unemployment by age from 1994 to 2004.

The issues of employment and unemployment for youth are relevant to pupils at school, even if they are going on to further study. There are many factors to be considered that underlie the changes in the workforce.

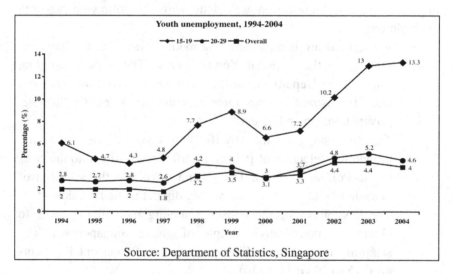

Figure 1. Singapore Youth Unemployment 1994-2004

The graphs in Figure 1 are based on those from Chart 2.12 in Chong & Chia (2006, p. 50).

Step 1: Getting started
Q: From the title, what is the general topic being examined?
Q: From the label on the vertical axis, how are the groups being compared?
Q: From the label on the horizontal axis, how are the groups being compared?
Q: What evidence is there that the information presented is reliable?
Q: To what does the legend refer?

Step 2: WHAT do the numbers mean?
Q: What is the precise meaning of the 6.1 on the left of the top graph?
Q: To what does the 4 on the right hand side on the lowest graph refer?
Q: Over what time frame are the data given?

Step 3: HOW do they differ?
Q: Look at the overall data from 1994 to 2004 and comment on the change in unemployment rates.
Q: Comment on the change in the rates for 15-19 year olds, 1994 to 2004.
Q: Comment on the change in the rates for 20-29 year olds, 1994 to 2004.
Q: Find the percentage change between the unemployment rate of 15-19 year olds in 1994 and those in 2004.

Step 4: WHERE are the differences?
Q: In which years are the differences between the unemployment rate for the 15-19 year olds and the overall unemployment rate greater than 3%?
Q: Compare unemployment rates for 15-19 year olds with those for 20-29 year olds.
Q: What do you notice if you compare 15-19 year olds with Overall unemployment? 20-29 year olds with Overall unemployment?

Step 5: WHY are there differences?
Q: Suggest reasons why the graphs for both 15-19 year olds and 20-29 year olds dip in 2000 and then start to rise again.
Q: Suggest reasons for differences between the unemployemt rates for the two youth groups.
Q: What social, environmental, cultural and economic factors might have influenced The youth unemployment in Singapore?

Through the use of the *Framework* with structured questions and tasks pupils can see that the unemployment rates for 15-19 year olds have changed more than those for 20-29 year olds. Pupils can reflect on what this means for them and their prospects for employment in full-time, part-time, temporary or casual jobs depending on their plans for the future. Pupils could compare this data with data from other countries where trends might be different as countries vary in terms of the extent of paid work that young people do and the availability of jobs for them.

5.2 *Activity two*

Analyse and interpret the data on world youth views on divorce

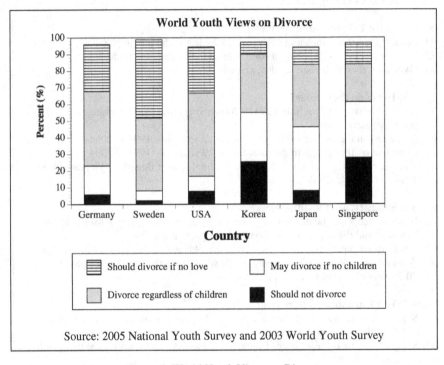

Figure 2. World Youth Views on Divorce

National Education in Singapore encourages pupils to examine the Singapore issues at home and reflect on how Singapore is similar to, and different from, other countries. The data plotted in Figure 2, based on Chart 5.5 in Chong & Chia (2006, p. 126), was collected through the 2005 Singapore Youth Survey and the 2003 World Youth Survey. Through the use of the *Framework* with structured questions and tasks pupils can read and interpret the graph and discuss the reasons for the

Developing Pupils' Analysis and Interpretation of Graphs and Tables 211

differences, taking into account the countries involved and the diversity of populations.

Step 1: Getting started
Q: From the title, what is the general topic being examined?
Q: From the label on the vertical axis, how are the groups being compared?
Q: From the label on the horizontal axis, how are the groups being compared?
Q: What evidence is there that the information presented is reliable?
Q: To what does the legend refer?

Step 2: WHAT do the numbers mean?
Q: What is the value of each division on the vertical scale?
Q: What percentage of youth in Singapore think that you *should not divorce*?
Q: What percentage of youth in the USA think that you *should not divorce*?
Q: What percentage of youth in Singapore say people *may divorce if no children*?

Step 3: HOW do they differ?
Q: Look at the heights of the columns. What do you notice? Why do you think this is?
Q: For Swedish youth order the 4 categories of youth views, ascending by percent.
Q: Write the six countries in ascending order according to the view that there should be *divorce regardless of the children*.
Q: Write the six countries in ascending order according to the view that there should be *divorce if no love*.

Step 4: WHERE are the differences?
Q: Compare Korea with Japan and comment on the main differences.
Q: Compare Singapore and Japan and comment on the main differences.
Q: Compare Germany, Sweden and the USA and comment on the main differences.

Step 5: WHY are there differences?
Q: Consider the category *should not divorce* for six countries. These vary from about 2% in Sweden to about 28% in Singapore. Suggest some reasons for these differences taking into account the major religions of the countries.
Q: Consider the category divorce regardless of children and why Germany's youth rate is almost twice as high as Singapore youth. What factors do you think might influence their views.

5.3 *Activity three*

Analyse and interpret the data on the benefits of social participation and social competencies among Singapore youth

The data in Table 2 are based on Table 4.2 in Chong & Chia (2006, p. 106). The context of this table is very important for today's youth and the data on the benefits of belonging to groups for developing social competencies can promote serious discussion in the classroom.

Table 2

Social participation and social competencies among Singapore youth

Benefits of social participation and social competencies among youths			
	Not involved in any group	Active in at least one group[2]	Difference
Multicultural orientation [3]			
Respect values and beliefs of other groups	17.9%	31.2%	13.3%
Knowledgeable about other groups	9.6%	14.8%	5.2%
Interpersonal relationships [3]			
Caring about other people's feelings	18.8%	35.1%	16.3%
Good at making friends	19.2%	34.5%	15.3%
Work well with others	14.6%	24.7%	10.1%
Leading a team of people	4.8%	15.0%	10.2%
Outward orientation [3]			
Public speaking	2.5%	7.7%	5.2%
Adapt to change	14.6%	24.9%	10.3%
Volunteering [4]			
To be actively involved in local volunteer work	34.8%	59.0%	24.2%
To be actively involved in overseas volunteer work	24.4%	38.3%	13.9%

Note: To illustrate the relationship between social participation and various social benefits, the table only displays a partial set of values ("very like me", "very important", and "somewhat important") for this set of variables. The chi-square test of association between this set of variables and involvement (incorporating the full set of values) are all statistically significant ($p < .01$).
[2] Active refers to those who are involved in a group on a weekly basis
[3] Percent indicating "very much like me"
[4] Percent indicating "very important" and "somewhat important" as a life goal
Source: 2005 National Youth Survey

Through the use of the *Framework* with structured steps pupils can analyse this data. The steps suggest considering the "difference" column in another way to help pupils make sense of the relative benefits of being part of social groups. This table looks relatively simple but pupils need to understand why the percents do not add to 100 and to consider the benefits of calculating differences. In addition, the calculation of ratios to make more sense of the data can lead to discussion of how additions to the table or the drawing of a graph can help.

Step 1: Getting started
Q: From the title, what is the general topic being examined?
Q: From the labels on the left column, how are the groups being compared?
Q: From the labels on the top row, how are the groups being compared?
Q: What evidence is there that the information presented is reliable?
Q: Look at the notes provided about the definitions of the superscripts.

Step 2: WHAT do the numbers mean?
Q: What is the precise meaning of the 17.9% in the first column?
Q: In *Interpersonal relationships*, which characteristic (on the left) is the highest percent in the *not involved in a group* category?
Q: In *Interpersonal relationships*, which characteristic (on the left) is the lowest percent in the *not involved in a group* category?

Step 3: HOW do they differ?
Q: In *Interpersonal relationships* for which characteristic (on the left) is the difference between *not being involved* and *being active* in a group: the greatest? the least?
Q: On graph paper, sketch a column or bar graph with *not involved* and *active* side by side for each category of *Interpersonal relationship* – comment on your graph.

Step 4: WHERE are the differences?
Q: For *Interpersonal relationships*, consider the ratio between the percents instead of the differences between them. Put in a fourth column of figures estimating (to the nearest one decimal point) the ratio of *active* to *not involved*. Find the biggest and the smallest ratios. Discuss.
Q: For which of *Multicultural orientation* or *Interpersonal relationships* does being *active* in a group make the most difference?
Q: What difference does being *active* in a group make to *Outward orientation*?

Step 5: WHY are there differences?
Q: What did you find that was as you would have expected?
Q: What social, environmental, cultural, economic or political factors might have influenced the young people's responses?

5.4 Activity four

Analyse and interpret data on Crime Cases recorded in Singapore 1998 to 2008

Table 3
Crime Cases Recorded in Singapore from 1998 to 2008

24.6 CRIME CASES RECORDED

Type of Offence	1998	2003	2004	2005	2006	2007	2008
	Number of Cases Recorded						
Overall Crime	40,090	30,282	30,623	37,039	33,263	32,796	32,412
Crimes Against Persons	3,904	4,292	4,109	4,608	4,103	4,113	4,354
Violent Property Crimes	1,028	1,053	925	1,190	1,004	1,027	960
Housebreaking and Related Crimes	1,770	1,411	1,299	1,551	1,201	926	889
Theft and Related Crimes	27,778	16,716	18,236	22,711	20,301	19,556	19,363
Commercial Crimes	2,717	3,230	3,111	3,389	3,159	3,565	3,490
Miscellaneous Crimes	2,893	3,580	2,944	3,644	3,495	3,609	3,356
	Per 100,000 Population						
Overall Crime Rate	1,021	736	735	870	756	715	670
Crimes Against Persons	99	104	99	108	93	90	90
Violent Property Crimes	26	26	22	28	23	22	20
Housebreaking and Related Crimes	45	34	31	36	27	20	18
Theft and Related Crimes	707	406	438	532	461	426	400
Commercial Crimes	69	78	75	79	72	78	72
Miscellaneous Crimes	74	87	71	85	79	19	69

Source: Police Intelligence Department

Developing Pupils' Analysis and Interpretation of Graphs and Tables 215

There are some changes in the number of crimes over time and pupils might compare these with the changes in the population of Singapore, (the total population and the residents of Singapore) which can be accessed on the Singapore Department of Statistics website. They can discuss how the reporting of crimes in itself changes and how the definitions of particular crimes can change. How is the peak in 2005 connected to different legislation or other factors? Through the use of the *Framework* with structured questions and tasks pupils can analyse this data.

Step 1: Getting started
Q: From the title, what is the general topic being examined?
Q: From the labels on the left column, how are the groups being compared? How many categories are there?
Q: From the labels on the top row, how are the groups being compared?
Q: What evidence is there that the information presented is reliable?
Q: Why does the table show numbers of cases as well as cases per 100,000 people?
Q: What is indicated by the title, *Crime Cases Recorded*?

Step 2: WHAT do the numbers mean?
Q: What is the meaning of the 40,090 in the first column?
Q: What is the meaning of the 707 in the first column?
Q: Which category has the highest number of cases in the time interval shown?
Q: Which category has the lowest percentage of cases in the time interval shown?
Q: In which year was the highest number of cases of *Commercial Crimes*?

Step 3: HOW do they differ?
Q: How does the number of cases of *Overall Crime* change from 1998 to 2008?
Q: How does the number of *Cases Against Persons* change from 1998 to 2008?

Step 4: WHERE are the differences?
Q: Look at the *Overall Crime* rate per 100,000 for the years from 1998 and 2003 and find the percentage change over that time. Find the percentage change from 2003 to 2008. Which is greater?
Q: Look at *Commercial Crimes* and *Miscellaneous Crimes* from 1998 to 2003 to see how they change. Are there any similarities?

Step 5: WHY are there differences?
Q: There was a substantial decrease in *Overall Crime* between 1997 and 2002. Why might this have been the case?
Q: For *Theft and Related Crimes*, the cases per 100, 000 decreased initially and then fluctuated. Can you think of any reasons why there was an increase in 2005?
Q: Reflect on changes in the recorded crime statistics in the light of any related social, economic and environmental, political or other factors.

6 Implications for the Mathematics Classroom

Pupils need to develop critical reading and thinking skills and the ability to put forward arguments with appropriate evidence. When analysing and interpreting data, or just reading a book or newspaper, they should automatically ask questions about the sizes of the samples and the ways in which they were they were collected. These skills are not easy to develop. Experience has shown that students can interpret graphs and tables with the help of structured questions or tasks. A challenge for teachers is to assist them to develop their own independent ways of working through the steps required for interpreting data. One approach is to give pupils a graph or table and ask them in groups to write suitable questions and discuss the answers. It is important to remember that pupils must be motivated, interested and need particular information if they are to independently engage in interpreting the data. One advantage of using media sources is that the data selected can be up-to-date and relevant to the real world of the pupils. In addition, there are other across-curricula resources that can provide situations where the pupils need to interpret the data and use the evidence to support their arguments.

7 Conclusion

The *Five Step Framework* provides a means for teachers to scaffold their students' thinking through the development of materials such as those illustrated in this chapter. The *Framework* permits teachers to use contemporary materials, of direct interest to students, addressing issues of significance to them personally as well as to their studies. Scaffolding is necessary in the initial stages but carefully selected examples of interest to students can be used to help pupils to see the kinds of intellectual work they need to engage in themselves and the kinds of expertise they need to develop to analyse and interpret statistical data.

References

Ben-Zvi, D., & Garfield, J. (2004). Statistical literacy, reasoning, and thinking: In D. Ben-Zvi & J. Garfield (Eds.), *The challenge of developing statistical literacy, reasoning and thinking.* (pp. 3-16). The Netherlands: Kluwer Academic Publishers.

Best, J. (2001). *Damned lies and statistics: Untangling numbers from the media, politicians and activists.* Berkeley: University of California Press.

Biggs, J. B., & Collis, K. F. (1982). *Evaluating the quality of learning: The SOLO taxonomy (Structure of the Observed Learning Outcome).* New York: Academic Press.

Chance, B., delMas, R. & Garfield, J. (2004). Reasoning about sampling distributions. In D. Ben-Zvi & J. Garfield (Eds), *The Challenge of Developing Statistical Literacy, Reasoning and Thinking* (pp. 295-323). Kluwer Academic Publishers: Dordrecht. The Netherlands.

Chapman, A., Kemp, M., & Kissane, B. (1990). Beyond the mathematics classroom: numeracy for learning. In S. Willis (Ed.), *Being numerate: What counts?* (pp. 91-118). Melbourne: The Australian Council for Educational Research.

Chong, H. K. & Chia, W. (2006). *Youth SG: The State of Youth in Singapore.* Singapore: National Youth Council.

Dewdney, A. K. (1993). *200% of nothing.* New York: John Wiley, Inc.

Ehrenberg, A. (1981). The problem of numeracy. *The American Statistician, 35*(2), 67-71.

Gal, I. (2003). Functional demands of statistical literacy: Ability to read press releases from statistical agencies. *International Statistical Institute 54th Session, Berlin, Germany.*

Kemp, M. (2005). *Developing critical numeracy at the tertiary level.* Digital thesis, Murdoch University. Retrieved from October 31, 2009, from http://wwwlib.murdoch.edu.au/adt/browse/search/?k

Kemp, M., & Bradley, B. (2006). Developing statistical literacy across the university curriculum. In D. Quinney, D. (Ed.), *International Conference on Teaching Mathematics*, Istanbul, Turkey, 2009. Retrieved May 31, 2009, from http://www.tmd.org.tr/ictm3/readme.html

Kemp, M., & Kissane, B. (1990). Numeracy: a tool for learning. In K. Milton & H. McCann (Eds.), *Mathematical turning points: Strategies for the 1990s* (pp. 366-375). Hobart: The Australian Association of Mathematics Teachers.

Koschat, M. (2005). A case for simple tables. *The American Statistician, 59*(1), 31-40.

Mosenthal, P.B., & Kirsch, I. (1998). A new measure for assessing document complexity: The PMOSE/IKIRSCH document readability formula. *Journal of Adolescent & Adult Literacy, 41*(8), 638-657.

Ministry of Education. (2006a). M*athematics syllabuses - Secondary.* Singapore: Author.

Ministry of Education. (2006b). *Social Studies syllabus - Lower Secondary Normal (Technical).* Singapore: Author.

Tufte, E. R. (1983). *The visual display of quantitative information.* Cheshire, Conn: Graphics.

Tukey, J. W. (1977). *Exploratory data analysis.* Reading, MA: Addison-Wesley.

Wainer, H. (1992). Understanding graphs and tables. *Educational Researcher, 21*(1), 14-23.

Watson, J. (2004). Developing reasoning about samples. In D. Ben-Zvi & J. Garfield (Eds), *The challenge of developing statistical literacy, reasoning and thinking.* (pp. 277-294). The Netherlands: Kluwer Academic Publishers.

Watson, J. (2006). Graphs — how best to represent the data. *Statistical Literacy at School: Growth and Goals* (pp. 55-96). New Jersey: Laurence Erlbaum Associates Inc.

Wong, K.Y. (2003). Mathematics-based national education: A framework for instruction. In S. K.S. Tan & C.B. Goh (Eds.), *Securing our future: Sourcebook for infusing National Education into the primary school curriculum* (pp. 117-130). Singapore: Pearson Education Asia.

Wu, Y., & Wong, K. Y. (2009) Understanding of statistical graphs among Singapore secondary students. In K. Y. Wong, P. Y. Lee, B. Kaur, P. Y. Foong & S. F. Ng (Eds), *Mathematics Education: The Singapore Journey* (pp. 227-243). Singapore: World Scientific.

Yearbook of Statistics Singapore (2009) Retrieved on November 18, 2009, from http://www.singstat.gov.sg/pubn/reference/yos09/statsT-miscellaneous.pdf.

Chapter 12

Theoretical Approaches and Examples for Modelling in Mathematics Education

Gabriele KAISER   Christoph LEDERICH   Verena RAU

The integration of applications and mathematical modelling in mathematics education is an internationally accepted goal emphasized in the scientific debate and school-based discussions. Although its necessity has been emphasised in large-scale international studies, the request to include applications and modelling in the mathematics classrooms is not new, and has a long tradition that goes back into the seventies of the twentieth century. The first part of this chapter focuses on the different theoretical perspectives about the integration of mathematical modelling in school developed in the last decades and emphasising rather different goals for the teaching of mathematical modelling in school. The second part is more example-oriented and describes several less complex modelling problems such as the distance of a ship from a Singaporean lighthouse. It also describes in detail an authentic modelling example, the chlorination of swimming pools which was carried out in a modelling week with students from upper secondary level at the University of Hamburg.

1 Introduction

The relevance of promoting applications and mathematical modelling in schools is widely accepted. For example the large international comparative OECD-study PISA emphasises developing the capacity of students to use mathematics in their present and future lives as goal of mathematics education. It means that students should understand the relevance of mathematics in everyday life, in our environment, and for the sciences.

This perception of the objectives of mathematics teaching has impact on the structuring of mathematics lessons. It is insufficient to simply impart competencies for applying mathematics only within the framework of the school curriculum. Instead more mathematics teaching should deal with examples from which

- students understand the relevance of mathematics in everyday life, in our environment, and for the sciences,
- students acquire competencies that enable them to solve real mathematics problems including problems in everyday life, in our environment, and in the sciences.

This demand for new ways of structuring mathematics teaching meets the goals for more reality-oriented mathematics education as postulated in many didactical positions since the middle or the end of the 20th century. It has been agreed that mathematics teaching should not be reduced to just reality based examples but that these should play a central role in education (for an overview of the historical debate see Kaiser-Messmer, 1986; the more recent debate is reflected in Blum et al., 2007). As a consequence it became clear that besides the application of standard mathematical procedures (such as applying well-known algorithms) in real world contexts and real world contexts serving as illustrations of mathematical concepts (e.g., usage of debts for the introduction of negative numbers), modelling problems as reality based contextual examples are increasingly important. In addition, it has been discussed that students should acquire competencies which enable them to solve real mathematical problems from different extra-mathematical domains. Within the current discussion it is stressed that it is not sufficient at all to deal with modelling examples in lessons but that the stimulation of modelling competencies through self-initiative is of central importance (see for a comprehensive longitudinal study Maass, 2004).

If one analyses many modelling examples discussed currently, it becomes obvious that, on the one hand different goals are connected with the introduction of modelling into mathematics lessons, while on the other hand the proposed modelling examples connected with these various goals are very different. In this chapter we want to discuss different theoretical conceptions connected with the teaching of mathematical modelling and describe a few modelling examples showing the variety of these approaches.

2 Theoretical Debate on Mathematical Modelling in Mathematics Education

The international discussion on mathematical modelling demonstrates that a homogeneous understanding of modelling and its epistemological backgrounds within the different groups working in the area of mathematical modelling in school does not exist. However, this is not a new situation at all and if we go back into the earlier debate on modelling we can find astonishing similarities.

Nearly twenty years ago, Kaiser-Messmer (1986, p. 83) showed in her analysis that within the applications and modelling discussion of that time various perspectives could be distinguished, for example, the following two main streams:

- A **pragmatic perspective,** focusing on utilitarian or pragmatic goals, the ability of learners to apply mathematics to solve practical problems. The work of Pollak (1969) can be regarded as a prototype of this perspective.
- A **scientific-humanistic perspective** which is oriented more towards mathematics as a science and humanistic ideals of education with focus on the ability of learners to create relations between mathematics and reality. The 'early' work of Freudenthal (1973) might be viewed as a prototype of this approach. Freudenthal, later changed his position, as he tended to take pragmatic aims more into consideration (see for example, 1981).

The various perspectives of the discussion vary strongly due to their aims concerning application and modelling. There are several classifications of the goals for the teaching of applications and modelling available. One such classification is as follows (see Kaiser, 1995, p. 69f, for a differentiation see Blum, 1996, p. 21f):

- **Pedagogical goals**: imparting abilities that enable students to understand central aspects of our world in a better way;
- **Psychological goals**: fostering and enhancement of the motivation and attitude of learners towards mathematics and mathematics teaching;
- **Subject-related goals**: structuring of learning processes, introduction of new mathematical concepts and methods including their illustration;
- **Science-related goals**: imparting a realistic image of mathematics as science, giving insight into the overlapping of mathematical and extra-mathematical considerations of the historical development of mathematics.

Meanwhile, the current discussion on modelling has developed further and become more differentiated. New perspectives can be identified which, as it became obvious from detailed analyses, emerged from the above described traditions or partly can be regarded as their continuations. The question, how the different goals connected to the introduction of mathematical modelling in school influence the kind of modelling examples proposed and the ways they are dealt with in school, has been strongly debated at the last three congresses of the European Society for Research in Mathematics Education (CERME 4, CERME5 and CERME6, in 2005, 2007 and 2009 respectively). Several classifications of the various historical and more recent modelling approaches were developed, and the first proposal was published by Kaiser and Sriraman (2006) and was modified later on by Kaiser et al. (2007). Based on these classification systems it is possible to distinguish the following theoretical approaches, which support the case for the inclusion of mathematical modelling in school. The suggested description of various approaches is based on recent analyses using literature mainly generated by ICMI (International Commission on Mathematics Instruction) and ICTMA (International Community of Teachers of Mathematical Modelling and Applications) and additional publications (see for example the reference list in the discussion document of the ICMI Study on applications and modelling in mathematics education, Blum et al., 2002, p. 279f).

The following classification distinguishes various perspectives within the discussion according to their central aims in connection with modelling and briefly describes the backgrounds to these perspectives as well as their connection to the initial perspectives. This ensures both a continuity for the present discussion as well as accumulates current perspectives coherently into the existing literature.

Name of the perspective	Central aims	Background	Selected recent protagonists
Realistic or applied modelling	Pragmatic-utilitarian goals, i.e.: solving real world problems, understanding of the real world, promotion of modelling competencies	Anglo-Saxon pragmatism and applied mathematics	Burkhardt, Haines, Kaiser
Model eliciting approach/ Contextual modelling	Subject-related and psychological goals, i.e. apply model elicited through solving the original problem to a new problem and solving word problems.	American problem solving debate	Lesh & Doerr, Sriraman
Educational modelling	Pedagogical and subject-related goals: a) Structuring of learning processes and its promotion b) Concept introduction and development c) Promotion of motivation and improvement of attitudes towards mathematics d) Promotion of critical understanding of modelling processes and models developed	Didactical theories and learning theories	Blum, Galbraith, Niss, Blomhøj, Maass, Stillman
Socio-critical	Pedagogical goals such as	Socio-critical	Barbosa,

and socio-cultural modelling	critical understanding of the surrounding world and of the models and the modelling process as overall goal, connected with recognition of cultural dependency of modelling approaches	approaches in political sociology, ethno-mathematics	Araújo
Epistemological or **theoretical modelling**	Theory-oriented goals, i.e. promotion of theory development through promotion of connections between modelling activities and mathematical activities, re-conceptualization of mathematics	Anthropological theory of didactics	Garcia

The following perspective can be described as a kind of **meta-perspective** focusing on research aims in contrast to the previous approaches:

Cognitive modelling	Restricted to research aims: a) analysis of cognitive processes taking place during modelling processes and understanding of these cognitive processes Psychological goals: b) promotion of mathematical thinking processes by using models as mental images or even physical pictures or by emphasizing modelling as a mental process such as abstraction or generalization	Cognitive psychology	Borromeo Ferri, Blum, Carreira

Figure 1. Classification of recent modelling approaches

First of all, it is remarkable that more recently, approaches from Roman language speaking countries were brought into the discussion on applications and modelling which start out from a more theory-related background (e.g. Garcia et al., 2006). As a consequence, the approaches of the epistemological perspective show a strong connection to the scientific-humanistic perspective mainly shaped by the early Freudenthal work. In his earlier work, Freudenthal (1973) distinguishes local and global mathematisation, and for global mathematisation the process of mathematising is regarded as part of the development of mathematical theory.

Likewise, analyses show that the approaches from the pragmatic perspective were sharpened further until they became the perspective of realistic modelling. For these kinds of approaches, authentic examples from industry and science play an important role. Modelling processes are carried out as a whole and not as partial processes, like applied mathematicians would do in practice. As a central characteristic of the realistic or applied perspective formulated by Burkhardt (2006), Haines and Crouch (2006) or Kaiser and Schwarz (2006) it can be stated that modelling is understood as activity to solve authentic problems and not as a development of mathematical theory.

Aside from these quasi polarising approaches, the realistic modelling and the epistemological modelling, there exists a continuation of integrative approaches within the perspective of educational modelling which puts the structuring of learning processes and fostering the understanding of concepts into the foreground of interest. The majority of approaches developed in the area on modelling can be classified under this perspective (see for example Blum, 1996; Blomhøj & Jensen, 2003; Galbraith et al., 2007; Niss, 1992; Maass, 2004). However, the approach of educational modelling may also be interpreted as a continuation of the scientific-humanistic approaches in its version formulated by Freudenthal in his later years and the continuation done by De Lange (1987) for whom real-world examples and their interrelations with mathematics become a central element for the structuring of teaching and learning mathematics.

Another perspective, the so-called socio-critical perspective, refers to socio-cultural dimensions of mathematics, which are closely associated with ethno-mathematics, promoted for example by D'Ambrosio (1999). This perspective emphasises the role of mathematics in society and claims the necessity to support critical thinking about the role of mathematics in society, about the role of and nature of mathematical models, and the function of mathematical modelling in society (e.g. Araújo, 2007; Barbosa, 2006).

The perspective of solving word problems - named contextual modelling - has a long tradition, especially in the American realm, but with the model eliciting perspective a theory-based perspective has been established which is clearly going far beyond problem solving at school. The model eliciting perspective is based on the premise that modelling research should take into account findings from the realm of psychological concept development to develop activities which motivate and naturally allow students to develop the mathematics needed to make sense of such situations (see Lesh & Doerr, 2003).

Within the discussion on applications and modelling, the approach of cognitive modelling, which examines modelling processes under a cognitive perspective, is new. This perspective aims to analyse various modelling processes with different types of modelling situations, varying in their degree of authenticity or mathematical complexity. One of the main goals is to reconstruct individual modelling routes or individual barriers and difficulties of students during their modelling activities. Researchers classified under this descriptive perspective of cognitive modelling such as Blum and Leiss (2007), Borromeo Ferri (2006) or Carreira (2001) might be found to be concerned about combining their normative approach on teaching applications and modelling within another perspective as well (see for example the approach developed by Blum et al., 2002, who is classified under the perspective educational modelling).

The above described framework for the description of the modelling debate allows insight into the origin of the different perspectives and its relations to the underlying background philosophy. It is made clear that the various approaches promoting applications and modelling in school or university teaching come from very different theoretical perspectives spanning the debate from ethno-mathematics to problem solving. They are characterised by different views on important aspects of applications and modelling, such as their views on goals and intentions of applications and modelling, which vary from promotion of a better understanding of the real world to the promotion of learning mathematical theory. Accordingly, their views on the role of the context are highly differentiated ranging from the call to authentic real world examples to more or less artificial, mathematically oriented examples. In addition, their perception of the modelling process is also highly different demanding a modelling cycle starting from real world problems and coming back to them or modelling processes which start from a real world problem, but lead to mathematical reflections and the development of new mathematical theory.

In the rest of the chapter various examples will be described, which show the variety of the different approaches. Within these examples the students are expected to carry out a modelling process, which can be done on the basis of the following ideal-typical procedure, as shown in Figure 2: A real world situation is the process' starting point. At the beginning the situation is idealised, that is, simplified or structured in order to get a real world model. Then this real world model is mathematised, that is translated into mathematics so that it leads to a mathematical model of the original situation. Mathematical considerations during the mathematical model produce mathematical results, which must be reinterpreted into the real situation. The adequacy of the results must be checked, that is, validated. In the case of an unsatisfactory problem solution, which happens quite frequently in practice, this process must be iterated.

Modelling process

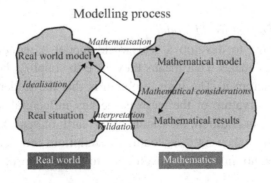

Figure 2. Modelling process (from Kaiser, 1995, p. 68 and Blum, 1996, p. 18)

3 Description of Various Modelling Examples

3.1 *Educationally oriented modelling examples*

(1) The Lighthouse Problem

The problem, how far a ship is away from a lighthouse, when the crew sees the fire of the lighthouse the first time, is a well-known sea navigation problem with high relevance in former times, before most ships were equipped with GPS. This problem is proposed by protagonists of the educational modelling perspective for the teaching of mathematical modelling in school (Blum, 2006) due to its mathematical richness and its easy accessibility and is adapted in the following to the local situation.

Horsburgh Light

Singapore's oldest lighthouse – the Horsburgh light - was built in 1851 and is 31 meters high. The lighthouse was intended to warn ships against approaching the coastline.

How far off the coast is a ship when the crew is able to see the light-fire for the very first time over the horizon? (Round off adequately).

Development of a real world model

The students have to develop a real world model based on different assumptions, they make. That means they have to simplify the situation, idealise and structure it. The main problem is that the students have to realise that the curvature of the earth is the key influential factor, they have to consider. In addition, the radius of the earth of approx. 6378.125 km needs to be taken into account. Assumptions concerning high visibility, that is, no mist and no swell need to be developed. The height of the observer, who might be between 1.7 to 1.9 m tall, need to be considered as well as the location of the observer, on sea level or on board in a lookout.

Development of a mathematical model
Next an appropriate mathematical model has to be developed: A first step can comprise the translation of the real model into a two-dimensional mathematical model describing the earth as a circle and then using the Pythagorean Theorem to calculate the required distance from the ship to the lighthouse.

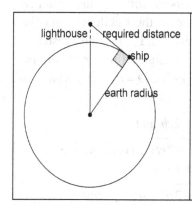

With the earth radius r = 6,371,000 m, the height of the lighthouse h = 31 m and d as required distance $r^2 + d^2 = (h+r)^2$.
Elementary transformations yield
$d = \sqrt{h^2 + 2rh}$.
So the required distance is about $d \approx 19,875 \, m = 19.875 \, km$.

Another attempt refers to the definition of the cosine, which can be used instead of the Theorem of Pythagoras.

One simplification refers to the fact, that the real distance of the ship is not equal to the length of the line of sight. However, using the formula of the section of a circumference in order to calculate the real distance of the ship does not yield significant differences between the distance on the earth and the line of sight

An extension of this simple model takes into account that the observer, who sees the lighthouse at first is not at the height of the waterline, but a few metres higher, for example being on a ship with 2.2 m height or even higher in a look-out. Furthermore we need to consider that the person has a height, for example 1.8 m tall. A possible approach uses the Pythagorean Theorem twice, firstly with the right-angled triangle from the geocentre to the top of the lighthouse to the boundary point, where the line of sight meets the sea surface.

The other right-angled triangle has the distance from the geocentre to the eyes of the person on the ship as the hypotenuse, which is r+H with H as height of the eyes of the observer, the distance from the eyes of the person to the boundary point as one side of the right-angled triangle and the earth radius as the other side.

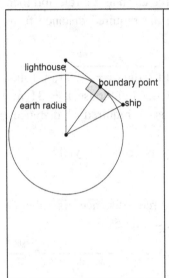

The requested distance D is the distance between the top of the lighthouse and the eyes of the person with d_1 the distance between the top of the lighthouse and the boundary point and d_2 the distance between the eye height of the observer and the boundary point: $D = d_1 + d_2$, which is calculated as follows:

$d_1 = \sqrt{h^2 + 2rh}$ and

$d_2 = \sqrt{H^2 + 2rH}$. Therefore the requested distance is:

$D = \sqrt{h^2 + 2rh} + \sqrt{H^2 + 2rH}$

$D \approx 27013m = 27.013km$

Interpretation and Validation
Next the results need to be interpreted and validated using knowledge from other sources. The results need to be transferred back to reality with the following questions:

- Does the result make sense? Is it realistic?
- Does it coincide with my own experience?
- Did I realise the assumptions correctly?

A validation of these two different results might include a comparison with known approximations from seafaring, which shows the higher adequacy of the second model to the real situation. In the

"Admiralty list of lights and fog signals" you can find the "geographical range table", which contains tables for the distance between a ship and a light-fire. The requested distance is, if the light-fire is 30 m high and the height of the eye of the observer is 4 m, 15.2 seamiles or 28.15 km.

Further explorations and extensions
The example of the lighthouse allows many interesting explorations, for example the development of an easier rule of thumb for the calculation of the distance of the ship and lighthouse. Referring to the easier case without considering the height of the ship and the person it uses the well-known physicists' argumentation, that small values such as h^2 are insignificant compared to $2rh$ and can be disregarded due to their small influence on the final result. That yields the following argumentation:

$$d = \sqrt{h^2 + 2rh} \approx \sqrt{2rh}.$$

The results are quite accurate, for example this rule of thumb yields as result for the first model approximately 20 km (for details see Blum, 2006).

It is possible to use this rule of thumb for further generalisations concerning the height of the lighthouse and the range of sight. One can easily see that the range of sight d is approximately proportional to $\sqrt{h}$ as height of the lighthouse that is the height of the lookout has to be quadruplicated to have twice the visibility.

Other continuations reflect on the reverse question, that is how far away is the horizon? That well-known problem is similar to the problem of the lighthouse and its solution is mathematically equivalent to the first elementary model. However, from a cognitive point of view, the real world model is much more difficult to develop, because the curvature and its central role is psychologically difficult to grasp. The development of the real model needs to consider that the visibility is dependent on height of observer, clearness of the air and so on. While developing the mathematical model the students should realise the usability of the Pythagorean Theorem, set up an equation, transform it and insert values. As mathematical result one notes that a child with 1.2 m height can see for 3.9 km and a tall man or woman of 1.8 height can see as far as 4.8

km. The last step contains an interpretation of the results and asks if, if the results make sense, if they are realistic, whether they relate to the own experience and so on.

One related example of current interest for students from lower secondary level might be the following:

	The Singapore Flyer is currently the world's largest Ferris wheel. It reaches a total height of 165 m. How far can you see when you are at the highest point of the Ferris wheel? (Round off your answer to the nearest whole kilometre.)

Using the models developed in the previous parts of the chapter yields an answer of approx. 46 km.

(2) Height of a Straw Bale Pile

The following example uses similar mathematical ideas as the first example but is based on a very different real world context. The problem, which stems from the DISUM project (http://www.disum.de/), is given to the students at school as follows:

	At the end of summer bales of straw can be admired on fields. The bales of straw in the picture were arranged in a way that five are in the bottom row, four are in the next one, then three, then two and one on top. Try to calculate the height of the whole pile of bales of straw as exact as possible.

A possible modelling process might look like this.

Step 1: Development of the real model
The bales of straw are built up as a pyramid. For this purpose five bales are in the bottom row and there are five rows on top of each other. Then it is important to know, that the bales of straw intertwine. Furthermore you can assume that every bale of straw is inherently stable.

As a comparative measure one can use the height of the woman in the picture. Considering the average height of Western young women one can assume that she is approximately 1.7 m tall and just as high as the bale of straw. So the diameter d of the bale of straw equals the height of the woman standing, therefore the radius of the bale of straw is 0.85 m.

Step 2: Development of the mathematical model
(a) *Model 1*. A very elementary model is to simply use the estimation of the size of the pictured woman for the estimation of the height of the straw pyramid. One can estimate, that the woman fits in there four times "one upon the other", hence the pile of bales is 6.8 m or approximately 7 m high.
(b) *Model 2*. In the second model more elaborated mathematics is used. Based on the Pythagorean Theorem the height of the straw bale pyramid can be calculated as shown in the picture that follows.

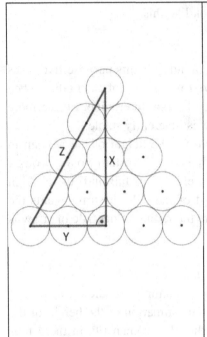

The side of the right-angled triangle y is as long as two diameters of a bale of straw: $y = 2d = 3.4\ m$. Moreover the hypotenuse z is as long as four diameters of a bale of straw: $z = 4d = 6.8\ m$. With these two results it is possible to compute the second side of the right-angled triangle x: $x = \sqrt{(6.8m)^2 - (3.4m)^2}$ the outcome of this is $x \approx 5.89\ m$.

Based on these results it is possible to calculate the overall height of the straw ball pyramid as follows:
$H = x + d = 7.59\ m$.
Rounding reasonably yields the overall height as $H = 7.6\ m$.

Step 3: Interpretation and Validation
The result seems to be reasonable. Both modelling approaches yield the result that the bale of straw pyramid is less than five times of the woman's height that is the pyramid of bales of straw is approximately 7.6 m high.

3.2 *Authentic complex modelling problems*

In Germany there exist several groups, who have implemented authentic complex modelling problems in mathematics teaching using the approach of so-called modelling weeks. These approaches refer to the modelling perspective "realistic or applied modelling" described at the beginning of the chapter. A central feature of these projects is the use of authentic examples, that is, problems that are suggested by applied mathematicians who work in industry who have met or tackled this problem within their working environment. The problems are only a little simplified and often no solution is known, either to the organisers of the project or to the problem poser. For example, the problem of unique identification of fingerprints is not yet solved satisfactorily; the machines do not work reliably, which creates concern at the widespread use of these machines for fingerprint identification. Quite often only a problematic situation is described and the students have to develop a question that can be solved – the development and description of the problem to be tackled is the most important and most ambitious part of a modelling process, mostly neglected in ordinary mathematics lessons. Another feature of the problems is their openness, which means that various problem definitions and solutions are possible depending on the views of the modellers.

The teaching-and-learning-process is characterised as autonomous, self-controlled learning that is the students decide their ways of tackling the problem and no fast intervention by the supervisors takes place (or should take place). Teachers are expected to offer only essential assistance, if mathematical means are needed or if the students are heading into a cul-de-sac. With this kind of teaching approach the students experience long phases of helplessness and insecurity which is an important aspect of modelling and a necessary phase within a modelling process.

Overview on modelling examples used so far:
Within the several modelling activities just described the following modelling problems have been tackled so far:
- Stock price forecast
- Premium in private health insurance
- Prediction of fishing quotas
- Optimal position of rescue helicopters in South Tyrol
- Radio-therapy planning for cancer patients
- Identification of fingerprints
- Pricing for internet booking of flights
- Price calculation of an internet café
- Traffic flow during the Soccer World Championship in 2006 in Hamburg
- Construction of an optimised time table of school

- Irrigation of a garden
- Development of ladybugs
- Chlorination of a swimming pool

4 Description of One Modelling Example in Detail

We will not describe very elaborate approaches but rather concentrate on the description of approaches that were developed by the students. The results that one of the student groups - named after the famous mathematician *d'Alembert*- achieved in the course of the week are to be briefly reconstructed in the following. They chose the problem of chlorinating a swimming pool: The company *Grundfos*, that markets pumping systems, is searching for a model describing the mixture of chemicals in a swimming pool. The concentration of urea and the number of bacteria in a swimming pool is to be adjusted by a systematic supply of chlorine. A mathematical model is needed for which describes the regulation of pool water quality via the chlorine concentration in swimming pools.

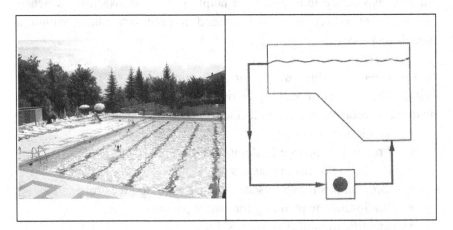

Figure 3. Outdoor swimming pool and sketch of the chlorination process

In the course of the week some students who were initially sceptical about the problem and mathematics in general abandoned their existing notions and prejudices and became significantly involved in the solution of the problem.

The results of the group can be interpreted as a single run through the modelling cycle.

4.1 *Real model*

At first, the different factors were listed that influence the quality of bathing water. The group agreed on examining the concentration of bacteria, urea and chlorine since different sources of information referred to varying factors. Those are exposed to the following influences:

Concentration of bacteria
1. New bacteria are added by the bathers.
2. These bacteria proliferate.
3. Chlorine kills the bacteria.

Concentration of urea
1. The concentration of urea is increased by the bathers.
2. Chlorine destroys the urea.

Concentration of chlorine
1. The concentration of chlorine decreases by the killing of bacteria.
2. By neutralising the urea, the concentration of chlorine decreases.
3. Since chlorine is gaseous at room temperature, it also evaporates from the water.
4. Chlorine is added via the pumping system.

It already becomes clear that specific modelling assumptions have to be made: Exclusively *concentrations* are considered. The amount of water in a swimming pool changes continuously in reality. However, the model does not take this into account because the considerations of a varying amount of water complicates the model to such an extent that it is mathematically difficult to handle. The volume of water is therefore considered to stay *constant*. The proliferation rate of bacteria in bathing water also depends on its temperature; therefore, the *water temperature* is considered to stay *constant* as well.

4.2 Mathematical model

(a) Simple model. Subsequently, the influences which are partly mutual and were mentioned above have to be transferred into a mathematical model. As is well known, the bacteria concentration B, the urea concentration H and the chlorine concentration C are considered. The urea and the chlorine concentration are given in [g m^{-3}] whereas the concentration of bacteria is measured in [KbE m^{-3}] as colony forming units. More precisely, the mathematical model consisted in examining the *modifications* of concentration beforehand. This was achieved by *functions of change f_B, f_H and f_C* which the students developed:

$$f_B = -v_B CB + z_B V^{-1} A + \lambda B \quad \left[KbE \cdot m^{-3} \min^{-1} \right] \qquad (0.1)$$

$$f_H = -v_H CH + z_H V^{-1} A \quad \left[g \cdot m^{-3} \cdot \min^{-1} \right] \qquad (0.2)$$

$$f_C = -n_B BC - n_H HC + Z_C - g \cdot C \quad \left[g \cdot m^{-3} \cdot \min^{-1} \right] \qquad (0.3)$$

Observing the modification of the concentration during time results in the units of the functions of change being [KbE m^{-3} min^{-1}] and [g m^{-3} min^{-1}]. The following listing reveals the meaning of designations:

- v_H is the urea destruction constant and v_B is the bacteria destruction constant. They indicate the amount of urea and bacteria respectively which is eliminated by the chlorine.
- z_H was defined as the constant of urea feeding and z_B was defined as the constant of bacteria feeding. They indicate the amount of urea and bacteria respectively that each bather supplies during a certain period of time.
- $A = A(t)$ depicts the number of bathers (of course varying over time).
- λ specifies the speed of bacteria reproduction.
- n_H and n_B respectively specify the amount of chlorine that is needed to neutralise the urea and eliminates the bacteria respectively.
- g specifies the amount of chlorine that evaporates.
- V specifies the volume of water.
- $Z_C = Z_C(t)$ specifies the feeding of chlorine concentration.

This is per se a system of ordinary differential equations. However, the designation of the *functions of change* entails that the group treated them as ordinary equations.

At this point, the example of the bacteria destruction constant v_B is to show exemplary which method was used in order to define the units of constants: The unit of the summand $v_B CB$ has to be [KbE m^{-3} min^{-1}]. Therefore, it must be calculated:

$$\left[\frac{KbE}{m^3 \cdot min} = \frac{m^3}{min \cdot g} \cdot \frac{g}{m^3} \cdot \frac{KbE}{m^3}\right]$$

Hence, the unit of the bacteria destruction constant v_B has to be [m^3 min^{-1}g^{-1}]. The exact values of the constants are shown in the following table.

constants	v_H	v_B	z_H	z_B
values of the constants	0.08 m^3 min^{-1}g^{-1}	0.453 m^3 min^{-1}g^{-1}	0.576 g min^{-1}	257 KbE min^{-1}

constants	λ	n_H	n_B	g
values of the constants	0.01155 min^{-1}	0.08 m^{3}min^{-1}g^{-1}	10^{-11}m^{3}min^{-1}KbE^{-1}	0.0025 min^{-1}

It is also necessary to take in account that the defined limits per DIN-norm are observed:

$$0.3 \frac{g}{m^3} \leq C \leq 0.6 \frac{g}{m^3} \tag{0.4}$$

$$H \leq 1 \frac{g}{m^3} \tag{0.5}$$

$$B \leq 100 \frac{KbE}{ml} = 10^8 \frac{KbE}{m^3} \tag{0.6}$$

But already the attempt to compare the different influences makes two problems clear:
1. The functions of change have varying units. Therefore, the values are not comparable.

2. The magnitudes of the single summands are also enormous as the values of the constants and the allowed bacteria concentration clearly show.

(b) Scaling or non-dimensionalisation. The model has to be scaled and non-dimensionalised respectively in order to eliminate these problems. The following operations were undertaken mainly by the advisors for this purpose: The urea and the chlorine concentration were scaled by the value 1g m^{-3}. The bacteria concentration was scaled by the valid bacteria concentration limit value of $B_g := 10^8$ KbE$\cdot m^{-3}$ in order to make its values comparable to those of the urea and chlorine concentration. In addition, the following new constants were introduced:

$$\tilde{n}_B := n_B \cdot B_g, \quad \tilde{z}_B := \frac{z_B}{V \cdot B_g}, \quad \tilde{z}_H := \frac{z_H}{V}$$

The non-dimensionalised mathematical model with the new terms of $\tilde{B}, \tilde{H}, \tilde{C}$ then reads as follows:

$$f_{\tilde{B}} = -v_B \tilde{C}\tilde{B} + \tilde{z}_B A + \lambda \tilde{B} \quad \left[\min^{-1}\right] \quad (0.7)$$

$$f_{\tilde{H}} = -v_H \tilde{C}\tilde{H} + \tilde{z}_H A \quad \left[\min^{-1}\right] \quad (0.8)$$

$$f_{\tilde{C}} = -\tilde{n}_B \tilde{B}\tilde{C} - \tilde{n}_H \tilde{H}\tilde{C} + \tilde{Z}_C - g \cdot \tilde{C} \quad \left[\min^{-1}\right] \quad (0.9)$$

The following table contains the new parameters which now all feature the unit min^{-1}. The water volume was thereby defined as 900 m^3. The functions of change hence became comparable and are all located in the same order of magnitude (between 0 and 1).

constants	v_H	v_B	$\tilde{z}_H$	$\tilde{z}_B$
values of the constants	0.08 min^{-1}	0.453 min^{-1}	$6.4 \cdot 10^{-4}$ min^{-1}	$2.86 \cdot 10^{-9}$ min^{-1}

constants	λ	$\tilde{n}_H$	$\tilde{n}_B$	g
values of the constants	0.01155 min^{-1}	0.08 min^{-1}	10^{-3} min^{-1}	$2.5 \cdot 10^{-3}$ min^{-1}

The supply of bacteria via the bathers can obviously be disregarded. The equation (0.7) then is: $f_{\tilde{B}} = \lambda \tilde{B} - v_B \tilde{C} \tilde{B}$

This function of change signifies on the one hand that the bacteria that exist in the water reproduce but on the other hand, the chlorine decreases the bacteria concentration. But which effect is prevailing? The *state of equilibrium* was regarded for this purpose. The state of equilibrium occurs if the bacteria concentration does not change, meaning that $f_{\tilde{B}} = 0$ is valid. A student who initially showed difficulties while working on the functions of change began to comprehend how to deal with them at this point:

> So, if I insert these values into the function of change, the result is negative. That means that the concentration of bacteria becomes smaller. So the destruction of bacteria by the chlorine is stronger than the reproduction of the bacteria. But if the result was positive, it would be the other way around and the concentration of chlorine would be too small. But when would this be the case? Oh, I can see when the function of change becomes 0, and then I have the limit.

This is exactly the case if $C_g = \lambda v_B^{-1} \approx 0.0255$. But this value is far below the compulsory minimum concentration of chlorine. If the concentration of chlorine lies within the permitted range, the bacteria concentration decreases rapidly. Thus, at no time does it influence the system essentially and was therefore disregarded altogether. The system then is simply:

$$f_{\tilde{H}} = -v_H \tilde{C}\tilde{H} + \tilde{z}_H A \quad \left[\min^{-1}\right] \quad (0.10)$$

$$f_{\tilde{C}} = -\tilde{n}_H \tilde{H}\tilde{C} + \tilde{Z}_C - g \cdot \tilde{C} \quad \left[\min^{-1}\right] \quad (0.11)$$

Now it was necessary to again examine equilibrium points of the system. If both $f_{\tilde{H}} = 0$ and $f_{\tilde{C}} = 0$ are valid, then the system is balanced and the chlorine concentration which is then added keeps the system stable. It was therefore recommended to plug in the exact permitted limited values as a load situation for the system – thus 0.6 for $\tilde{C}$ and 1 for $\tilde{H}$ in order to then discover the number of visitors and amount of chlorine feeding that is needed to keep the system stable. Insertion and calculation provides:

$$A = 75 \quad \tilde{Z}_C = 0.0495 \left[g \cdot m^3 \cdot \min^{-1}\right]$$

In this context it should be noted that the 75 people count for the preconditioned water volume $V = 900 m^3$. If a different volume is conditioned at the outset, the number of people who may at most be in the pool can easily be calculated so that the chlorine and urea concentration barely lie within the permitted scope.

Finally, this was done for several swimming pools in Hamburg. The result is shown in the following table.

Name of the swimming pool	Volume $[m^3]$	Number of people	Feeding of chlorine [g min^{-1}]
Kaifu	286	23	14.157
Kaifu	337.5	28	16.7
Kaifu	1800	150	89.1
Kaifu	2760	230	136.125
Bondenwald	1041	86	51.53
Bondenwald	157	13	7.77
Bondenwald	2400	200	118.8
Indoor and outdoor pool Billstedt	1041	86	51.53
Indoor and outdoor pool Billstedt	95	7	4.7
Indoor and outdoor pool Billstedt	1890	157	93.555
Indoor and outdoor pool Billstedt	2760	230	136.62
Alsterschwimmhalle	3500	291	173.25
Alsterschwimmhalle	195.75	16	9.69
Alsterschwimmhalle	130	10	6.435

It is important to note that the values that were calculated altogether lie not in a realistic order of magnitude. The "Alsterschwimmhalle" for example uses 200 kg chlorine for all pools, altogether 3800 m^3, in 5.5 weeks. Therefore the feeding of chlorine is 3.65 g min^{-1}. Another example is the swimming pool "Bondenwald". The first pool, which is mentioned in the table, has a pump, which can add a maximum of $400 g h^{-1} = 6.67 g \min^{-1}$ to the pool. But on average the pump put only 2.83 g min^{-1} in the pool. The calculated results are definitely too high. One reason is, that we calculate with the maximum number of guests, but there never are so many guests. In addition the technician put only the necessary portion of chlorine in the pool, but the students calculated with the maximum, which is allowed.

Lastly there are many other influences, which are neglected by the modelling approaches of the students.

5 Conclusions

From the evaluation results it is obvious that authentic problems are not too difficult for most students, most of them did not feel overtaxed. Although at some time during the modelling week the students had experienced helplessness, lack of orientation and insecurity, these impressions did not change their positive judgement of the whole modelling experience. If one looks in detail at the views of the students, it becomes clear that most of them experienced the modelling examples as difficult but not too difficult. Most students described the problems as being reasonably structured. It was a consensus that the problems were realistic and interesting.

Furthermore the students felt they had achieved high learning outcomes, which reflect all the goals connected with modelling, ranging from psychological goals such as motivation to meta-aspects such as promoting working attitudes to pedagogical goals, namely enhancing understanding of the world around us. The strong plea by the students for the inclusion of these kinds of examples in normal mathematics lessons supports our position, that these authentic complex modelling examples are not too difficult for students.

We therefore call for including these kinds of problems in ordinary mathematics lessons, clearly not everyday, but on a regular basis. The realisation of this ambitious goal to include these kinds of examples in ordinary mathematics lessons will obviously create many difficulties to the usual structuring of teaching in portions of 45 minutes. It will be necessary to carry out curricular changes, i.e. shortening of the subject-related requirements, specialised topics need to be abandoned as well as many routine algorithms and exercises. But if we really want to convince our students that mathematics is important for everyday life for everybody, we need to work with these kinds of examples on a regular basis. We can conclude that at least for students these authentic modelling problems are not too difficult.

References

Araújo, J.L. (2007). Modelling and the critical use of mathematics. In C.P. Haines, P. Galbraith, W. Blum, & S. Khan (Eds.), *Mathematical Modelling (ICTMA 12): Education, Engineering and Economics* (pp. 187-194). Chichester: Horwood.

Barbosa, J.C. (2006). Mathematical Modelling in classroom: a critical and discursive perspective. *ZDM – The International Journal on Mathematics Education*, 38(3), 293-301.

Blomhøj, M. & Jensen, T.H. (2003). Developing mathematical modelling competence: conceptual clarification and educational planning. *Teaching Mathematics and its Applications*, 22(3), 123-139.

Blum, W. (1996). Anwendungsbezüge im Mathematikunterricht – Trends und Perspektiven. In G. Kadunz, H. Kautschitsch, G. Ossimitz & E. Schneider (Eds.), *Trends und Perspektiven* (pp. 15-38). Wien: Hölder-Pichler-Tempsky.

Blum, W. (2006). Modellierungsaufgaben im Mathematikunterricht – Herausforderung für Schüler und Lehrer. In A. Buechter, H. Humenberger, S. Hußmann & S. Prediger (Eds.), *Realitätsnaher Mathematikunterricht – vom Fach aus und für die Praxis* (pp. 8-24). Hildesheim: Franzbecker.

Blum, W. et al. (2002). ICMI Study 14: Applications and Modelling in Mathematics Education – Discussion Document. *Educational Studies in Mathematics*, 22, 37-68.

Blum, W., Galbraith, P.L., Henn, H.-W. & Niss, M. (Eds.) (2007). *Modelling and Applications in Mathematics Education. The 14th ICMI Study.* New York: Springer.

Blum, W., & Leiss, D. (2007). How do students and teachers deal with modelling problems? In C.P. Haines, P. Galbraith, W. Blum, & S. Khan (Eds.), *Mathematical Modelling (ICTMA 12): Education, Engineering and Economics* (pp. 222-231). Chichester: Horwood.

Borromeo Ferri, R. (2006). Theoretical and empirical differentiations of phases in the modelling process. *ZDM – The International Journal on Mathematics Education*, 38 (2), 86-95.

Burkhardt, H. (2006). Functional mathematics and teaching modelling. In C.P. Haines, P. Galbraith, W. Blum, & S. Khan (Eds.), *Mathematical Modelling (ICTMA 12): Education, Engineering and Economics* (pp. 177-186). Chichester: Horwood.

Carreira, S. (2001). The mountain is the utility – on the metaphorical nature of mathematical models. In J.F. Matos, W. Blum, K. Houston, & S. P. Carreira (Eds.), *Modelling and Mathematics Education: ICTMA9: Applications in Science and Technology* (pp. 15-29). Chichester: Horwood.

D'Ambrosio, U. (1999) Literacy, matheracy and technocracy: a trivium for today. *Mathematical thinking and learning*, 1 (2), 131-153.

De Lange, J. (1987), *Mathematics – insight and meaning*. Utrecht: Rijksuniversiteit Utrecht.

Freudenthal, H. (1973). *Mathematics as an Educational Task*. Dordrecht: Reidel.

Freudenthal, H. (1981). Mathematik, die uns angeht. *Mathematiklehrer*, 2, 3-5.

Galbraith, P.; Stillman, G.; Brown, J., & Edwards, I. (2007). Facilitating middle secondary modelling competencies. In C.P. Haines, P. Galbraith, W. Blum, & S. Khan (Eds.), *Mathematical Modelling (ICTMA 12): Education, Engineering and Economics* (pp. 130-141). Chichester: Horwood.

García, F.J., Gascón, J., Ruiz Higueras, L., & Bosch, M. (2006). Mathematical modelling as a tool for the connection of school mathematics. *ZDM – The International Journal on Mathematics Education*, 38(3), 226-246.

Haines, C.R., & Crouch, R.M. (2006). Getting to grips with real world contexts: Developing research in mathematical modelling. M. Bosch (Ed.), *Proceedings of the Fourth European Society for Research in Mathematics Education* (pp. 1634-1644). Universitat Ramon Llull.

Kaiser-Messmer, G. (1986). *Anwendungen im Mathematikunterricht. Vol. 1 – Theoretische Konzeptionen. Vol. 2 – Empirische Untersuchungen*. Bad Salzdetfurth: Franzbecker.

Kaiser, G. (1995). Realitätsbezüge im Mathematikunterricht – Ein Überblick über die aktuelle und historische Diskussion. In G. Graumann, T. Jahnke, G. Kaiser, & J. Meyer. (Eds.), *Materialien für einen realitätsbezogenen Mathematikunterricht* (pp. 66-84). Bad Salzdetfurth: Franzbecker.

Kaiser, G., & Schwarz, B. (2006). Mathematical modelling as bridge between school and university. *ZDM – The International Journal on Mathematics Education*, 38(2), 196-208.

Kaiser, G., & Sriraman, B. (2006). A global survey of international perspectives on modelling in mathematics education. *ZDM - The International Journal on Mathematics Education*, 38(3), 302-310.

Kaiser, G., Sriraman, B., Blomhøj, M., & Garcia, F.J. (2007). Report from the Working Group Modelling and Applications - Differentiating Perspectives and Delineating Commonalities. In D. Pitta-Pantazi, & G. Philippou (Eds.), *Proceedings of the Fifth Congress of the European Society for Research in Mathematics Education* (pp. 2035-2041). Larnaca: University of Cyprus.

Lesh, R., & Doerr, H. (Eds.) (2003). Beyond constructivism: *models and modelling perspectives on mathematics problem solving, learning, and teaching*. Mahwah: Lawrence Erlbaum Associates.

Maass, K. (2004). *Mathematisches Modellieren im Unterricht. Ergebnisse einer empirischen Studie.* Hildesheim, Franzbecker.

Niss, M. (1992). *Applications and modelling in school mathematics – directions for future development.* Roskilde, IMFUFA Roskilde Universitetscenter.

Pollak, H. (1969). How can we teach applications of mathematics. *Educational Studies in Mathematics*, 2, 393-404.

Chapter 13

Mathematical Modelling in the Singapore Secondary School Mathematics Curriculum

Gayatri BALAKRISHNAN   YEN Yeen Peng   Esther GOH Lung Eng

Over the last decade, there has been increased attention on mathematical modelling and its impact on the learning of mathematics at the school level. Applications and modelling were introduced in the framework of the Singapore school mathematics curriculum in 2003. This chapter draws out the four elements of the mathematical modelling cycle, viz. *Mathematisation*, *Working with mathematics*, *Interpretation*, and *Reflection*. These elements describe the activities that students will engage in during the modelling process. The purpose of this chapter is to help secondary school mathematics teachers understand the nature of the mathematical modelling process and familiarise them with the four elements. An example of a mathematical modelling task has been used to illustrate the mathematical modelling process.

1 Introduction

Over the last decade, there has been increased attention on mathematical modelling and its impact on the learning of mathematics at the school level (Blum, Galbraith, Henn & Niss, 2002). In mathematical modelling, students solve real-world problems mathematically. Mathematical modelling is a cyclic process (English, 2007) of translating a real-world problem into a mathematical problem by formulating and solving a mathematical model, interpreting and evaluating the solution in the real-world context, and refining or improving the model if the solution is not acceptable. The emphasis in mathematical modelling at the school level is on the process rather than on the product.

Mathematical modelling allows students to connect classroom mathematics to the real world, showing the applicability of mathematical ideas (Zbiek & Conner, 2006; Stillman, 2009). Given a real-world problem, students need to understand the real-world situation and make assumptions in order to devise a mathematical method to tackle the problem. Thus, mathematical modelling deepens students' understanding and enriches students' learning of mathematics. When students work in groups to tackle the problem, they also develop important 21st century skills such as collaborative learning skills and metacognitive skills (Tanner & Jones, 2002; McClure & Sircar, 2008).

Applications and modelling were introduced in the framework of the Singapore mathematics curriculum in 2003 to highlight their importance in mathematics learning so as to meet the challenges of the 21st century (Soh, 2005; Ministry of Education, 2009). The framework is shown in Figure 1.

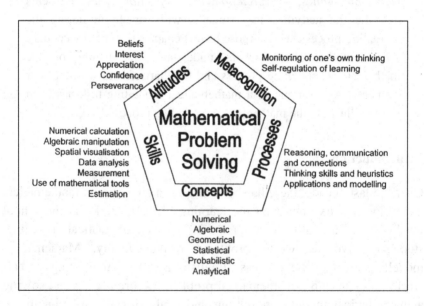

Figure 1. Singapore Mathematics Framework (Ministry of Education, 2006a)

2 Mathematical Modelling in the Singapore School Mathematics Curriculum

As shown in Figure 1, the framework of the Singapore mathematics curriculum has mathematical problem solving as its central focus. Five inter-related components, namely, *Concepts*, *Skills*, *Processes*, *Attitudes*, and *Metacognition*, contribute to the development of mathematical problem-solving ability. The range of problems referred to in the framework includes non-routine, open-ended and real-world problems.

Applications and modelling play a vital role in the development of mathematical understanding and competencies. It is important that students apply mathematical problem-solving skills and reasoning skills to tackle a variety of problems, including real-world problems.

Mathematical modelling is the process of formulating and improving a mathematical model to represent and solve real-world problems. Through mathematical modelling, students learn to use a variety of data representations, and to select and apply appropriate mathematical methods and tools in solving real-world problems. The opportunity to deal with real data and use mathematical tools for data analysis should be a part of the learning of mathematics at all levels.

3 Elements of the Mathematical Modelling Cycle

Figure 2 is a diagrammatic representation of the mathematical modelling cycle. It comprises four elements viz. *Mathematisation, Working with mathematics, Interpretation*, and *Reflection*. These elements describe the activities that students will engage in during the modelling process. The purpose of this chapter is to help secondary school mathematics teachers understand the nature of the mathematical modelling process and familiarise them with the four elements. The modelling process starts with a real-world problem - a problem that arises from a real-world situation using real-world data.

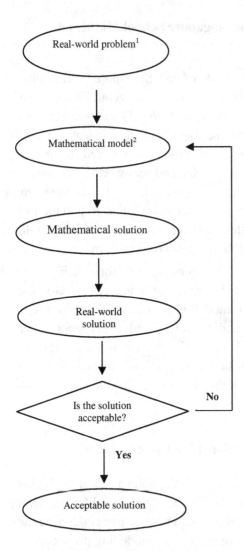

Mathematisation
- Understand the problem
- Make assumptions to simplify the problem
- Represent the problem in mathematical form

Working with mathematics
- Solve the mathematics problem using mathematical methods and tools including ICT

Interpretation
- Interpret the mathematical solution within the context of the original problem

Reflection
- Review the assumptions and the limitations of the mathematical model and its solution
- Review the mathematics methods and tools used
- Improve on the mathematical model

Figure 2. The Mathematical Modelling Process (version 2009)

---

[1] Real-world problem: A problem that arises from a real-world situation using real-world data.
[2] Mathematical model: A mathematical representation of a problem (e.g. scale drawing, diagram, table, graph, function, equation)

Mathematisation is the process of translating the real-world problem into a mathematical one by formulating a mathematical model. This requires students to understand the problem which may be open-ended and complex. They need to examine the given information, make suitable assumptions and simplify the problem to one that can be solved. In the process, they will identify the mathematics concepts and variables, represent the problem in a mathematical form and set up a mathematical model such as a drawing, graph, function or a set of equations.

Working with mathematics requires students to select and use appropriate methods and tools to solve the problem, after having formulated the problem mathematically. The students may make use of computer software to help them analyse data, carry out tedious computations and solve the problem without advanced mathematics methods. The end product at this stage is a mathematical solution.

Interpretation connects the mathematical solution back to the real world. The students make sense of the solution in the real-world context.

Reflection is the metacognitive aspect of modelling. It involves the review of assumptions and limitations of the model, the mathematical methods and tools used in solving the problem. This could lead to an improvement in the model as well as the solution. For example, in reviewing the model and mathematics that they use, students may consider using another method or tool if the solution obtained is not acceptable.

4 Modelling Tasks

Good modelling tasks allow students to experience the entire modelling process. They are real-world problems which may be open-ended, unstructured and complex. Students should also see meaning in solving these problems. As the task is open-ended, the students may arrive at different solutions. The following task has been designed to illustrate the modelling process.

4.1 Example of a modelling task

In a New Town, there is a park and an empty plot of land. People can cycle through the park, park their bicycles and then walk to a MRT station. To make the park safe, the Town Council wants to build a cycling track and a bicycle bay where the commuters can park their bicycles before they walk to the station. Working in groups, prepare a proposal for the Town Council to consider.

This is a real-world problem and students can work in groups on the problem. The teacher could use photographs to show the current situation or a possible scenario in the past that has led to the construction of cycling tracks and bicycle kiosks. This modelling task provides some insights into how the location of facilities in a town can be determined — the design objective and the mathematics required. The elements of the modelling process shown in Figure 2 are explained in detail.

I Mathematisation

1. To understand the problem, students may use a diagram or draw a sketch to represent the situation.
2. Students need to make some assumptions and collect information such as
   - the dimensions of the park and the plot of land;
   - the average walking speed and cycling speed;
   - the speed limit when cycling in a park;
   - the design objective (e.g. shortest path, minimum time, etc.).

3. They need to identify the mathematics concepts involved before formulating the model. In this case, the problem involves the concept of distance, speed and time.
4. The formulation of the mathematical model would depend on the assumptions made. For example, if the students used the minimum time as the design objective, the expression for time would be as follows:

$$\text{Total time taken, T} = \frac{\text{Distance travelled in the park}}{\text{cycling speed}}$$
$$+ \frac{\text{Distance travelled on the plot of land}}{\text{walking speed}}$$

Figure 3 shows a possible sketch to represent the diagram with the following assumptions:
- The shapes for the park and the plot of land are rectangular
- The dimensions & speed are based on information from the internet
- Design objective: Minimum time

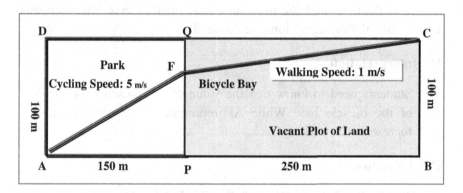

Figure 3. Sketch to represent the real world problem

II Working with Mathematics

5. Students could use dynamic geometry software such as GSP to move the point F to different positions and measure the distances, tabulate their observations and determine the position of F that gives the minimum value of time.
6. Students could formulate an expression for the distances using Pythagoras' theorem and formulate the function for time.

$$\text{Total time taken, T} = \frac{\sqrt{150^2 + (100-x)^2}}{5} + \frac{\sqrt{250^2 + x^2}}{1}, \text{ where } x$$

represents the distance QF

Students could use graphing software to draw the graph for this expression to help them find the value of x, corresponding to the minimum value of T. Upper secondary students could also use their knowledge of calculus to solve the problem.

Technology plays an important role in the modelling process. The affordance of technology makes the task accessible to a diverse group of learners at different levels. Technology enables students to deal with the task even though they may not have some requisite knowledge such as Pythagoras' theorem or the knowledge of calculus to find the solution. Dynamic geometry software like Geometer's Sketchpad (GSP) allows students to investigate the problem assuming different shapes and dimensions of the park and the plot of land. The graphing software allows students to find the minimum value and observe what happens when some of these conditions are changed.

III Interpretation

7. Students need to know that the value of x represents the location of the bicycle bay. While AF represents the cycling track, CF represents the walking track.

IV Reflection

8. Having found the location of the bicycle bay, the students need to check whether it is feasible to build the bicycle bay at the location on the actual site and consider if there are any other constraints.
9. Students could review their assumptions.
    - Could the park be circular in shape? What if the park and the plot of land are irregular polygons? Would their method be applicable in all cases? How do town planners actually do it?
    - Is their design objective practical? They could survey people to find out whether people prefer a solution that minimises time or they would prefer to walk the least distance instead.
10. This task can actually be done using different methods. Students could think about the mathematics that they have used to solve the problem and whether there are more powerful softwares that can be used to solve this problem.

11. Students could be encouraged to make connections of their learning with other subjects, for example, the use Snell's Law in Physics whereby the faster path obeys the laws of refraction.

The modelling task allows students to go beyond finding the mathematical solution and requires students to examine the feasibility of the solution in the context of the real world. By engaging in the mathematical modelling process, students develop skills such as communication skills, collaborative skills, thinking skills, metacognitive skills and ICT skills.

5 Students' Blockages during the Modelling Process and their Implications

There are potential blockages when students engage in mathematical modelling (Galbraith & Stillman, 2006; Maaβ, 2006). Students could have difficulties understanding the problem, making assumptions and identifying the key variables required to formulate the mathematical model. They are also limited by their mathematics knowledge and ability to choose an appropriate method to solve the problem and to interpret the solution. The understanding of these blockages provides insights into students' thinking and learning as they go through the modelling process.

It is important that teachers provide scaffolding to guide students as they go through the various elements of the mathematical modelling process. Teachers need to identify the necessary prerequisite knowledge and skills required for students to formulate the mathematical model and solve it. Students can work in groups to tackle the problem collaboratively, using technology where possible. Technology empowers students to work on problems which would otherwise require more advanced mathematics.

6 Conclusion

In this chapter, a mathematical modelling task has been used to illustrate the four elements of the mathematical modelling process, which is a rich learning process for students. The task allows students to work collaboratively to deal with real-world problems, which are open-ended and authentic. Students should develop awareness of the mathematical modelling process and experience all or some of the elements of the process during their secondary school years.

For a start, teachers can introduce some elements of the mathematical modelling process during mathematics lessons by using relevant examples of real-world problems from textbooks. They may adapt existing resources such as the Strategies for Active and Independent Learning Package (SAIL) (Ministry of Education, 2006b), Performance Assessment in Mathematics (Fan, 2008), and Programme for International Student Assessment (PISA) tasks (Organisation for Economic Co-operation and Development, 2006). Teachers can also use modelling tasks to allow students to experience the entire mathematical modelling process. Students may work in groups on these tasks, with the use of technology, within or outside the curriculum time.

Teachers should be familiar with the mathematical modelling process in order to facilitate students' learning as they go through the mathematical modelling process. Clarity of these four elements described above is a good starting point for teachers wanting to bring mathematical modelling activities into their classrooms for the first time.

References

Blum, W., Galbraith, P. L., Henn, H.-W., & Niss, M. (2002). ICME Study 14: Applications and modelling in mathematics education – discussion document. *Educational Studies in Mathematics, 51*(12), 149-171.

English, L. D. (2007). Interdisciplinary modelling in the primary mathematics curriculum. In J. Watson, & K. Beswick (Eds.), *Proceedings of the 30th annual conference of the mathematics education research group of Australasia* (pp. 275-284). Australia: MERGA Inc.

Fan, L. (Ed.). (2008). *Performance assessment in mathematics: Concepts, methods, and examples from research and practices in Singapore classrooms.* Singapore: Ministry of Education.

Galbraith, P., & Stillman, G. (2006). A framework for identifying student blockages during transitions in the modelling process. *ZDM, 38*(2), 143-162.

Maaβ, K. (2006). What are modelling competencies? *ZDM, 38*(2), 113-142.

McClure, R. H., & Sircar, S. (2008). Quantitative literacy of undergraduate business students in the 21st Century. *Journal of Education for Business,* 83(6), 369-374.

Ministry of Education (MOE). (2006a). *Mathematics syllabus (Secondary).* Singapore: Author.

Ministry of Education (MOE). (2006b). *Strategies for Active and Independent Learning (SAIL).* Singapore: Author.

Ministry of Education (MOE). (2009). *The Singapore model method for learning mathematics.* Singapore: Panpac Education.

Organisation for Economic Co-operation and Development (OECD). (2006). *Programme for International Student Assessment (PISA) released items — Mathematics.* Retrieved May 4, 2009, from *http://www.oecd.org/dataoecd/14/10/38709418.pdf*

Soh, C. K. (2005, Nov). *An overview of mathematics education in Singapore.* Paper presented at the 1st International Mathematics Curriculum Conference, Chicago, USA.

Stillman, G. (2009, Jun 4). *Implementing applications and modelling in secondary school: Issues for teaching and learning.* Keynote address delivered at the Mathematics Teachers Conference, Singapore.

Tanner, H., & Jones, S. (2002). Assessing children's mathematical thinking in practical modelling situations. *Teaching Mathematics and its application, 21*(4), 145-159.

Zbiek, R. M., & Conner, A. (2006). Beyond motivation: Exploring mathematical modelling as a context for deepening students' understanding of curricular mathematics. *Educational Studies in Mathematics, 63,* 89-112.

Chapter 14

Making Decisions with Mathematics

TOH Tin Lam

Despite the lack of a universal agreement on the definition of "mathematical modelling", one can broadly perceive modelling as the process of harnessing the power of mathematics to solve real-world problems. This chapter discusses how decision mathematics fits into the criterion of mathematical modelling, which can be carried out in a secondary school mathematics classroom, thereby fitting into the problem solving framework of the Singapore mathematics curriculum. Suitable tasks from decision mathematics to engage students in modelling, and the implications of modelling in the curriculum are also discussed.

1 Introduction

Mathematical modelling has a long history and has appeared prominently in many tertiary mathematics courses. In spite of this, there is a lack of universally agreed definition of the term "mathematical modelling" (Ang, 2006). Some researchers view all applications of mathematics as mathematical modelling (Burghes, 1980). Ang (2009) argued that mathematical modelling can be thought of "as a process in which there is a sequence of tasks carried out with a view to obtain a reasonable mathematical representation of the real world". Some mathematics educators define mathematical modelling as the process of "using the power of mathematics to solve real-world problems" (Hebborn, Parramore & Stephens, 1997, p. 42).

Despite the many differences in opinions among researchers on the term "mathematical modelling", one common feature that occurs prominently throughout the diverse opinion of mathematical modelling can be identified: mathematical modelling involves real-life problems.

The idea of using real-life problems in teaching and learning school mathematics is not new. In fact, there is a worldwide trend that school mathematics brings real-life situations into the mathematics classroom (Lesh & English, 2005). It has also been shown that activities in real-life situations could help students to see the relevance of mathematics and feel its power of applicability in the real world, thereby developing students' interest in learning mathematics (Correa, Perry, Sims, Miller, & Fang, 2008; Morita, 1999; Mitchell, 1994; Toh, 2009). McAlevery and Sullivan (2001) argued that "student are best motivated by exposure to real applications, problem cases and projects." This process of relating school mathematics content to real-life situations could be seen as the forerunner of introducing mathematical modelling into the school curriculum.

Typical mathematics textbooks include sections on "real-life applications" of the particular mathematics concepts. Ang (2009) observed that most of these "real-life application" problems have very neat answers and solutions. Tasks with very tidy state with perfect answers do not resemble authentic tasks and such questions can hardly convince students of real-life applications of mathematics (Ang, 2009). Real-life tasks are usually "untidy" and involve inter-disciplinary knowledge sets and skills to complete the tasks. This aspect is lacking in the typical mathematics textbooks.

2. Mathematical Modelling in the Curriculum Framework of Mathematical Problem Solving

2.1 *Importance and success of mathematical modelling*

The importance of mathematical modelling is recognized worldwide by mathematicians and mathematics educators. Smith and Wood (2001,

p. 48) argued that "this is an effective way of introducing mathematical concepts to engineering students, whose main interest in mathematics is often limited to its usefulness in their future profession." It is even recognized that one of the most important tasks of the 21st century mathematicians is to recognize repeated patterns in different real-life problems and propose a mathematical model in order to explain them (Vasco, 1999, cited in Martinez-Luacles, 2005, p. 194).

Modelling activities are usually immersed in a set of activities that do not rely too heavily on direct instructional methods and procedures. Thus, learners become engaged in authentic tasks that are mathematical in nature (Carrejo & Marshall, 2007).

Mathematical modelling has recently been introduced to children as young as nine-year-olds (e.g. English & Watters, 2005a; 2005b). Modelling has helped young students develop important mathematical ideas and processes that they normally would not meet in the early school mathematics curriculum. Through modelling activities, students can be trained to elicit important mathematical ideas that are embedded within meaningful real-world contexts (English & Watters, 2005a).

Studies have also shown that introducing modelling into the mathematics curriculum has also resulted in teachers modifying their way of teaching, making their teaching of mathematics more related to real-life problems (Martinez-Luacles, 2005).

2.2 *Mathematical problem solving and modelling in the Singapore school mathematics curriculum*

Mathematical modelling requires a different set of skills, knowledge and attitudes from teachers than problem solving (Martinez-Luaces, 2005). How, then, does modelling fit into the problem solving framework of the Singapore mathematics curriculum framework (Figure 1)?

An observant reader would have noticed the component "Applications and Modelling" classified under the arm "Processes" in Figure 1. Thus, application and modelling should be valued in the same way that mathematical reasoning and communication, thinking skills and heuristics are valued in the curriculum.

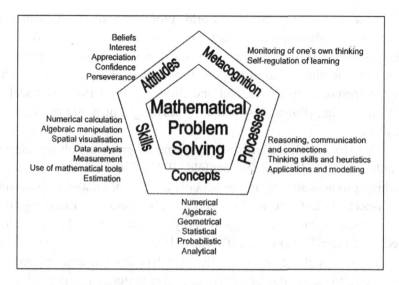

Figure 1. Framework of the Singapore school mathematics curriculum
(Ministry of Education, 2006)

In the author's opinion, modelling could help to achieve the goal of problem solving objective of the Singapore mathematics curriculum.

A mathematical problem is a situation in which an individual or group is called upon to perform a task for which there is no readily accessible algorithm which determines completely the method of solution. Furthermore, the individual or the group has the interest or desire to perform the task (Lester, 1978; 1980).

Real-world problems are usually "authentic tasks" for which the solutions to the problems are not immediately obtainable. Furthermore, students are best motivated to solve authentic tasks (McAlevery & Sullivan, 2001). Thus, a real-world problem fits into Lester's definition of a mathematical problem.

Solving a real-world problem requires the initial process of translating the essential elements of the problem into a mathematical model (Hebborn et al., 1997). The process of translation includes a thorough understanding of the constraints of the real-world problem, and the parameters that one has to operate. This reinforces the first stage of Polya's problem solving model (Polya, 1945): understand the problem.

After translating the real-world problem into a mathematical model, the problem solver needs to use appropriate strategies to devise "algorithms" to solve the newly translated problem. The "algorithms" should be applicable to solve other related problems as well. This stage corresponds to the second and third stages of Polya's model of problem solving (Polya, 1945): that of devising a plan and carrying out the plan.

Meta-cognition is a major determinant of problem-solving success or failure in an individual (Schoenfeld, 1985). Much of the difficulty in teaching problem solving in our schools arises from under-emphasizing this aspect (Toh, Quek & Tay, 2008). The aspect of meta-cognition corresponds to the fourth stage of Polya's model of problem solving: check and extend the problem (Polya, 1945).

In the modelling process, after one has devised and applied the algorithms to solve the problem, it is instructional for the solver to reflect on their solutions. Extending beyond the single problem, the problem solver might even be curious to generalize and extend the method to other related problems: "Are there other problems that this method can be applied?"

In conclusion, mathematical modelling, if properly carried out in mathematics classrooms, could help teachers reinforce the various aspects of mathematical problem solving of the mathematics curriculum.

2.3 *Making decisions with mathematics and discrete mathematics*

The author proposes the use of decision-making activities to involve school students in authentic tasks involving mathematical modelling.

Many real-world problems involve making choices based on mathematical computations or mathematical processes or "heuristics". After the problem solver understands the problem, he or she translates the real-world problem into a mathematical one, devises plausible plans to solve the problem and obtains more than one potentially correct solutions to the problem (Espinel Febles & Antequera Guerra, 2008). The decision maker will have to assess the possible implications of each possible solution in relation to the objective of the problem. Even in the event of obtaining one correct solution, the problem solver may decide whether the solution is an optimal one, by revisiting the solution steps or using estimations or bounds.

Based on the above description, one would be quick to realize that many problems associated with discrete mathematics are suitable for mathematical modelling tasks, and for helping students associate the world around them with mathematics (Grenier, 2008).

Discrete mathematics has seen the most progress in the twentieth century. The National Council of Teachers of Mathematics (NCTM) and the American Mathematical Society (AMS) have promoted its incorporation into the mathematics curricula. Many notions of discrete mathematics, such as optimization techniques, are very useful in the world of mathematics today (Kenny & Hirsch, 1991; Rosenstein, Franzblau & Robert, 1997).

However, it should be emphasized that discrete mathematics should NOT be introduced into the school mathematics curriculum merely as additional mathematics content to be taught to school students. Rather, discrete mathematics provides many useful ideas for teachers on how mathematical modelling can be carried out in schools.

3 Activities on Decision Making

The author has worked with groups of teachers and students on the feasibility of introducing mathematical modeling into the secondary school mathematics classroom. In this section, three problems that were used by the author are introduced.

Problem 1 depicts a scenario in a typical Singaporean context about a typical housing estate in Singapore. The context is close to the heart of every Singaporean, but the content, which involves identifying the minimum spanning tree in a weighted graph by either Prim's or Kruskal's algorithm (e.g. Hebborn et al., 1997, pp. 43-58), is beyond the school mathematics curriculum.

Teachers should NOT force their students to memorise the algorithms, but encourage the students to devise their own plan on how to handle this problem, and to reflect over their own solutions of the problem. Through these processes, the students' problem solving skills is reinforced while appreciating mathematical modeling.

Problem 1. Following an effort to upgrade one of the Singapore housing estates, the Town Council is planning to build sheltered pathways to link all the blocks of the flats shown in Figure 2. The cost of building the sheltered pathways (in thousands of Dollars) across each pair of the blocks is also shown. The Town Council wants to keep the building cost as low as possible, with all the blocks of the flats connected. How would you advise them to build the sheltered pathways?

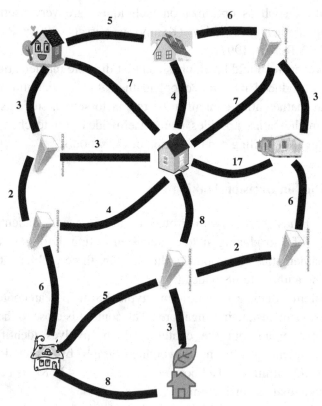

Figure 2. Diagram showing the connected paths among the various housing blocks in an estate

In the task of Problem 1, teachers could begin by emphasizing to their students the importance of understanding the problem – the first stage of Polya's model of problem solving. Following that, the teacher could then guide the students to model Figure 2 into a simple weighted graph – a much simplified diagram consisting of only vertices and edges as in Figure 3. Modelling here involves filtering off "redundant" information and leaving the essential information in the simplest possible form.

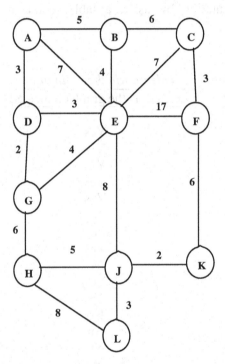

Figure 3. Modelling of the diagram of the housing estate using a directed graph

Solving Problem 1 is equivalent to solving the mathematical problem of Figure 3 by eliminating as many edges from the graphs as possible and ensuring that all the vertices in Figure 3 are connected, with the minimum sum of the values of all the connected edges. Using the language of discrete mathematics, this process involves finding the minimum spanning tree of the graph of Figure 3. In this process, students are introduced to the process of translating a real-world problem into a mathematical problem, thereby solving the mathematical problem.

To find the minimal spanning tree of the graph in Figure 3 by eliminating as many heavy edges as possible, the importance of "systematizing" the steps of elimination could be highlighted. It would be preferable to eliminate the heaviest edges from the diagram, with the graph remaining connected. Here, students would appreciate that the problem solving heuristic of "systematic listing", especially with systematic presentation by using a table, could help to solve this problem.

Table 1
Listing all the edges to decide on which to eliminate

Edge	Weight	If deleted, is the graph connected?	Decision
EF	17	Yes	Delete
EJ	8	Yes	Delete
HL	8	Yes	Delete
AE	7	Yes	Delete
CE	7	Yes	Delete
GH	6	Yes	Delete
FK	6	No	Stay
BC	6	No	Stay
AB	5	Yes	Delete
HJ	5	No	Stay
EG	4	Yes	Delete
BE	4	No	Stay
AD	3	No	Stay
DE	3	No	Stay
CF	3	No	Stay
JL	3	No	Stay
DG	2	No	Stay
JK	2	No	Stay

Based on the process of eliminating the edges in decreasing weight, Table 1 was obtained, and which could be translated into the minimum spanning tree in Figure 4.

After the students have obtained an initial solution as in Figure 4, it is important for the teachers to challenge their students to convince that their solution is the best possible one, or to check other possible solutions. For example, in Table 1, if the edge FK is eliminated instead of GH (both have weight 6), what will be the total weight of the new minimum spanning tree compared with the previous one? This process of reflection encourages the students to develop the habit of re-thinking the accuracy of their initial solution and to re-examine their algorithm of solving a mathematics problem.

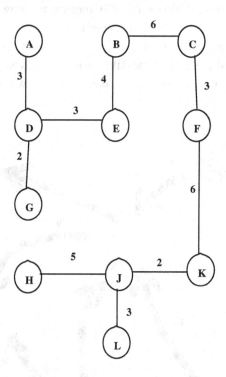

Figure 4. Minimum spanning tree of the graph in Figure 3

The teacher could take a further step to invite their students to consider other related real-world situations which they have encountered that appear to fit into the above category of problems, and which can be solved by the above method that they have "devised".

In the author's discussion of the above problem with practicing classroom teachers, they agreed that students should be exposed to more of such real-world problems, with teachers guiding the students through the mathematical processes instead of providing them with immediate solutions, methods or hints. They acknowledged the difficulties that both teachers and students might initially encounter. Moreover, teachers should be sufficiently proficient in higher level mathematics content knowledge.

Problem 2. The distance between any two stations in a section of the zoological garden in multiples of 100 metres is shown in Figure 5. Suppose the routes are only unidirectional, as indicated by the arrows. Albert wants to travel from the Tiger Enclosure to the Kangaroo Enclosure within the shortest possible route. Advise Albert on the shortest possible route that he should take. Also, show how you arrive at your solution carefully.

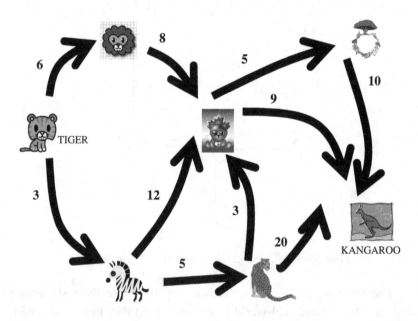

Figure 5. Floor plan of a zoo showing the various animal enclosures

The students should be guided to translate this diagram into one involving a directed graph (Figure 6). The teacher could lead students to find the shortest path from the Tiger Enclosure to the Kangaroo

Enclosure. Students could discover that the method by listing out all possible paths might be extremely laborious, leaving them the curiosity and discover an "easier" and more systematic method to solve the problem.

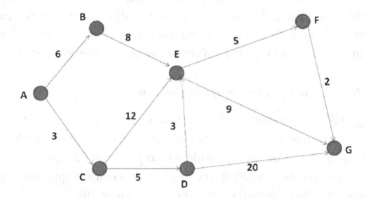

Figure 6. Modelling of the floor plan of the zoo in Figure 4 using a directed graph

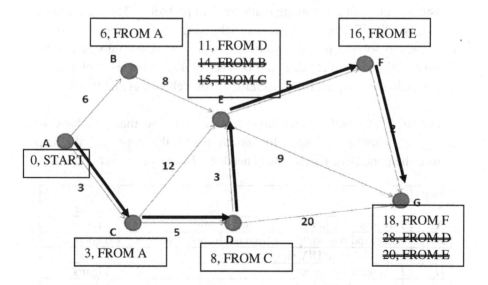

Figure 7. Identifying the shortest path from A to G through appropriate labeling

Through teachers' facilitation, students might actually devise a systematic procedure of locating the shortest path, for example, by giving a meaningful label to each vertex representing each enclosure. Ultimately, the label of the vertex representing the Kangaroo enclosure yields the value of the shortest distance (Figure 7).

This problem can be easily solved by Dijkstra's algorithm (e.g. Hebborn et al., 1997), which involves labeling of vertices. Without introducing the standard Dijkstra's algorithm directly, teachers could guide the students to devise and systematize their own "algorithms" and lead them to appreciate the value of the technique of labeling vertices.

3.1 *Modelling and constraints of the real world*

"One egg takes 3 minutes to boil. How many minutes will 10 eggs take?" Students' answer to this question reflects the extent to which they link mathematics to real-life. A student who provides 30 minutes as the answer obviously has applied the pure mathematical procedure of multiplication without regard for the constraints of reality.

Through modelling, students become more conscious of the constraints of reality in solving mathematical problems. They are trained to apply significant mathematical thinking and problem solving heuristics in solving real-life problems with their constraints of reality. In this section, a further example is discussed (labeled as Problem 3). This problem is adapted from Exercise 4A of Hebborn et al. (1997).

Problem 3. Consider the following task. Imagine that you have to completely change the lounge by completing all the steps with the help of three helpers. How much time is needed? Justify your answer.

Step	Task	Time needed
A	Buy new curtain from renovation centre	4 hours
B	Select and buy wallpaper from renovation centre	6 hours
C	Buy paint from DIY shop	1 hour
D	Remove the carpets	2 hours
E	Remove old curtains	4 hours
F	Paint the woodwork	4 hours
G	Hang the new wallpaper	4 hours
H	Hang the new curtains	4 hours
I	Replace the carpets	2 hours

In order to complete the above set of tasks, one needs to be aware of the precedence of each step and the available resources. These steps lead the students to develop the concept of "scheduling" – more than one task can be scheduled provided the precedence is met, and resources are available.

In this problem, students need to be conscious that certain steps need precedence to get started – a realistic reflection of many real-life tasks. Steps A, B, C, D and E need no precedence, step F needs C, D and E as precedence, G needs B and F as precedence, H needs A and G as precedence, I needs H as precedence.

After identifying the precedence, students would attempt to think of the representation of the most ideal way of scheduling the tasks among the three available workers of completing the task in the shortest possible time. Here, a good problem solving heuristic would be to devise a systematic diagram to represent this scheduling process as clearly as possible. An example of such a diagram is shown in Figure 8.

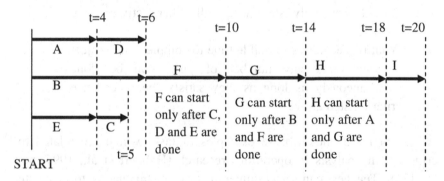

Figure 8. An example of a schedule to complete the entire project

By the choice of a suitable diagram (e.g. Figure 8), a reflective student would observe that in order to complete this task, it would be good to employ more helpers at the beginning of the project. However, after step F begins, it is not economical to employ more than one helper since all the subsequent steps require the preceding tasks as precedence.

Once the students are familiar with the idea of scheduling, they can be guided to develop their own "algorithms" to solve such scheduling problems. Teachers can then introduce a more general question for students to abstract the algorithm they have devised to solve scheduling problems to handle a more general setting as in Generalised Problem 3(a) below, which is a generalisation of Problem 3 in Hobbert et al. (1997, p. 102).

> **Generalised Problem 3(a).** A project consists of activities A, B, C, D and E. The time taken to complete each of these tasks is 10, 20, 25, 15 and 30 minutes respectively. However, they must satisfy the following conditions:
> Tasks A and B are independent of each other.
> Only when Task A is completed can Task C start.
> Task D requires Task A to be completed before it can start.
> Task E can only start when all other activities are finished.
> What is the shortest possible time to complete the project, assuming that any number of tasks can be done simultaneously as long as they satisfy the precedence criteria listed above?

Problems 3 and 3(a) are examples of tasks which are related to critical path analysis in operations research (Hobbert et al., 1997, pp. 99–131). Teachers can select numerous tasks of this nature to inculcate in their students a sense of application of mathematical processes in the real-world. Through the discussion of how problems 1, 2 and 3 can be solved, it could be seen that through such proposed modelling activities, students' problem solving strategies are reinforced.

4 Conclusion

The problems discussed in the preceding section show that mathematical modelling may not necessarily involve sophisticated technological tools, or sophisticated knowledge of mathematical theories, which are beyond the reach of school students.

In conclusion, it should be reiterated that mathematical modelling does not conflict with mathematical problem solving. For example, the problems selected in the above discussion can be used to develop students' logical thinking, concretize their devised procedures in solving the problems related to real world situations. Not only is there no conflict between mathematical modelling and problem solving, modelling helps to reinforce and re-enact the entire process of mathematical problem solving, which is the centre of the Singapore mathematics curriculum.

There are students who lack intrinsic interest in mathematics. For this group of students, introducing mathematical modelling into the curriculum may lead them to see the relevance of mathematics in solving real-life problems, thereby developing their interest in the subject.

Studies have also shown that significantly many teachers have modified their way of teaching in their mathematics classrooms as a result of being exposed to experiences in mathematical modelling.

However, there is no easy answer as to how to promote mathematical modelling into everyday teaching of mathematics. Ultimately, the ideal situation would be for teachers to lead students to see modelling "as a way of life" (Houston, 2001), thereby developing an intrinsic interest in mathematical problem solving.

If mathematical modelling were to be introduced as a key component into the Singapore mathematics curriculum, teachers' capacity in higher level mathematics knowledge must be further enhanced. It can be seen from the problems raised in this chapter that many problems that land themselves neatly into modelling and which are not necessarily involved with sophisticated technological tools are selected from higher level mathematics, especially discrete mathematics.

References

Ang, K. C. (2006). Teaching mathematical modelling in Singapore schools. *The Mathematiccs Educator, 6*(1), 63–75.

Ang, K. C. (2009). Mathematical modelling and real life problem solving. In B. Kaur, B.H. Yeap, & M., Kapur. (Eds), *Mathematical Problem Solving: Association of Mathematics Educators Yearbook 2009* (pp. 159–182). Singapore: World-Scientific.

Burghes, D. (1980). Mathematical modelling: A positive direction for the teaching of applications of mathematics at school. *Educational Studies in Mathematics, 11*, 113–131.

Carrejo, D. J., & Marshall, J. (2007). What is mathematical modelling? Exploring prospective teachers' use of experiments to connect mathematics to the study of motion. *Mathematics Education Research Journal, 19*(1), 45–76.

Correa, C. A., Perry, M., Sims. L. M., Miller, K. F., & Fang, G. (2008). Connected and culturally embedded beliefs: Chinese and US teachers talk about how their students best learn mathematics. *Teaching and Teacher Education: An International Journal of Research and Studies, 24* (1), 140–153.

English, L. D., & Watters, J. J. (2005a). Mathematical modelling with 9-year-olds. In H. L. Chick, & J. L. Vincent (Eds.), *Proceedings of the 29th Conference of the International group for the Psychology of Mathematics Education* (Vol. 2, pp. 297–304). Melbourne: PME.

English, L. D., & Watters, J. J. (2005b). Mathematical modelling in the early school years. *Mathematics Education Research Journal, 16*(3), 58–79.

Espinel Febles, M. C., & Antequera Guerra, A. T. (2008). The decision-making as a school activity. *Paper presented at the 11th International Congress of Mathematics Education (TSG15)*, Monterrey, Mexique, July 4–11, 2008.

Grenier, D. (2008). Some specific concepts and tools of discrete mathematics. *Paper presented at the 11th International Congress of Mathematics Education (TSG15)*, Monterrey, Mexique, July 4–11, 2008.

Hebborn, J., Parramore, K., & Stephens, J. (1997). *Decision mathematics 1*. Bath: The Bath Press.

Houston, K. (2001). Teaching modelling as a way of life. *Quaestiones Mathematicae*, Supp. 1, 105–113.

Kenney, M. J., & Hirsch, C. R. (Eds.) (1991). *Discrete Mathematics across the curriculum K-12*. Inc, Reston, Virginia: NCTM.

Lesh, R., & English, L.D. (2005). Trend in the evolution of models and modelling perspectives on mathematical learning and problem solving. *ZDM, 37*(6), 487–489.

Lester, F. K. (1978). Mathematical problem solving in the elementary school: Some educational and psychological considerations. In L. L. Hatfield & D. A. Bradbard

(Eds.), *Mathematical problem solving: Papers from a research workshop* (pp. 53–88). Columbus, Ohio: ERIC/SMEAC.

Lester, F. K. (1980). Research on mathematical problem. In R. J. Shumway (Ed.), *Research in Mathematics Education* (pp. 286–323). Reston, Virginia: NCTM.

Martinez-Luaces, V. E. (2005). Engaging secondary school and university teachers in modelling: some experiences in South American countries. *International Journal of Mathematical Education in Science and Technology, 36*(2/3), 193–205.

McAlevey, L.G., & Sullivan, J.C. (2001). Making statistics more effective for business? *International Journal of Mathematics Education in Science and Technology, 32* (5), 425–438.

Ministry of Education. (2006). *A Guide to Teaching and Learning of O-Level Mathematics 2007*. Singapore: Author.

Mitchell, M. (1994). Enhancing situational interest in the mathematics classroom. *Research Report for a meeting*. New Jersey.

Morita, J. G. (1999). Capture and recapture your students' interest in statistics. *Mathematics Teaching in the Middle School, 4*(6), 412–418.

Polya, G. (1945). *How to solve it*. Princeton: Princeton University Press.

Rosentein, J. G., Franzblau, D. S., & Robert, F. S. (1997) (Eds.), *Discrete Mathematics in the Schools*, DIMACS Series in Discrete Mathematics Computer Science, Volume 36, Providence, RK: American Mathematical Society (AMS).

Schoenfeld, A. H. (1985). *Mathematics Problem Solving*. Orlando, Florida: Academic Press.

Smith, G., & Wood, L. (2001). Motivating engineering students to study mathematics. *Mathematics Teacher*, Supp. 1, 47–56.

Toh, T. L. (2009). Arousing students' curiosity and mathematical problem solving. In B. Kaur, B. H. Yeap, & M. Kapur (Eds.), *Mathematical Problem Solving: Association of Mathematics Educators Yearbook 2009* (pp. 241–262). Singapore: World-Scientific.

Toh, T. L., Quek, K. S., & Tay, E. G. (2008). Mathematical problem solving - a new paradigm. In J. Vincent, R. Pierce, & J. Dowsey (Eds.), *Connected Maths* (pp. 356–365). Victoria: The Mathematical Association of Victoria.

Chapter 15

The Cable Drum – Description of a Challenging Mathematical Modelling Example and a Few Experiences

Frank FÖRSTER   Gabriele KAISER

In this chapter the mathematical modelling example of a 'cable drum' will be described. This deals with the question of the length of the cable fitting on a large cable-drum used for roadwork. The example is suitable for students from grades 9 to 12 (15 to 18 year old students) and allows many different ways to deal with it. In relation to the mathematical proficiency of the students and their modelling competencies, simple modelling approaches mainly based on the circumference of a circle as well as more complex modelling approaches using a helix are accessible. In the chapter, the experiences with this example in a case study are described, i.e. the modelling approaches developed by the students and their errors are discussed in detail. Finally, the modelling process contains a few dead ends, which allow interesting reflections concerning the relation of mathematics and the real world.

1 The Modelling Problem

They are not common but nearly everybody has seen them at least once: huge cable drums that even require the trailer of a flat bed truck for transportation. These cable drums are applied whenever electric or telephone cables are inserted into the ground by special machinery during road constructions (see Figure 1) and they are used all around the world.

The Cable Drum

Figure 1. Usage of a cable drum in road work (copyright ©Helga Lade)

The question, we are going to deal with in the proposed modelling example, is simply as follows: *How many metres of cable fit on this cable drum?* (see Figure 2). The example was first published in 2002 (Förster & Herget, 2002) and was described with results from empirical research in Förster (2008).

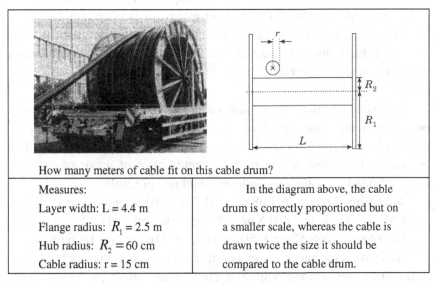

Figure 2. Worksheet about the cable drum

The dimensions of such a cable drum are indeed impressive: The so-called *layer width L* that amounts to 4.4 m is already wider than an ordinary road. In addition there are the flanges of appr. 10-15 cm width. The *flange radius* R_1 is 2.5 m and the cable drum therefore reaches a total height of 5 metres. The *hub radius* R_2 is 60 cm and the cable radius r is 15 cm.

It is advisable to form your own estimations about the length of the cable and think about options to calculate the cable length before you continue reading.

2 Different Approaches to Model the Problem and First Experiences

The problem is markedly well suited for 8th and 10th grade students in the context of circle calculations. It gains attraction in higher grades, because more mathematical means are available and lower secondary level knowledge has to be reactivated. The teaching sequence was frequently tested in German 9th and 10th grade classes (in different types of school) as well as with first-year university students and those studying to become a teacher. We offer as few impulses as possible having the experience that the question is self-explanatory. The problem is especially suited for working in small groups; it is open and diverging not only because various approaches are conceivable but also because they are actually used by the students. A natural differentiation happens of its own accord depending on particular capabilities of the students – we will return to this later.

Although the problem of the cable drum seems to be clear and precise and the basic approach via circle calculation seems to easily accessible, the students arrive at a wealth of different results. For instance, an example lesson in a 10th grade class of upper proficiency level unfolded the following 'colourful bouquet of results':

442.68 m; 712 m; 791.68 m; 791.7 m; 791.92 m; 795 m; 799 m; 820 m; 870.6 m; 870.8 m; 900 m; 904 m; 941.72 m; 942 m; 1015.68 m; 1068.57 m; 1150 m; 1162 m; 1163 m und 1266.69 m.

According to our experience, it rarely occurs that only a few different results emerge within a class, often some working groups already present several conclusions. Certainly, some values are immediately detected to be an approximation or rounding of another value and of course, students as well miscalculate occasionally. Nevertheless, the variety of results is surprising. Almost every result is obtained by rational modelling and is seldom based on 'blind trying'. The models that are presented in the following are either adequate models in order to answer the question on the worksheet or models that lead to an inadequate real model, because of fundamental and structural misinterpretations of the reality. Models, which include 'transmission errors' from the real model to the mathematical model, will not be considered at first. We will return to this later. There are illustrations selected for each particular example. These illustrations have been outlined in such a way or similarly by the modellers during the development of the real model. The actual calculations are only indicated but can easily be reconstructed.

2.1 *Model M1 ("Many-circle-model")*

The model that is by far most frequently used, and is considered at least in the beginning of almost every processing of the problem, comes from the description of the cable as layered annuli. Andrea's group working on the task described their attempt as follows:

"After initial deliberations we came to the conclusion that the number of cable circles on the roll had to be calculated (14 × 6). This result rendered the possibility to calculate each circumference but we had to bear in mind that the radius always increases by 0.3 m. Finally, we added the resulting values."

The overall problem 'How many metres of cable fit on the cable drum altogether?' is thus separated in four sub-problems (see Figures 1 and 2):

- How many (cable-) circles fit next to each other?
- How many layers of cable fit on the drum?
- How many metres of cable fit in one single circle of the first (second, third, …) layer?
- How many metres of cable fit in the first (second, third, …) layer of the drum?

Within the real model the helix form of windings around the hub within one layer of cable is disregarded as well as the transitions from one to the next layer (see Figure 3). Actually, this is an underlying – generally very conscious – modelling approach: the cable wrappings that actually adjoin one another are replaced by individual separated circles – those are relatively easy to calculate. For this, the cable is mostly layered one above the other. Sketches or references in the text show that this is consciously seen as a simplifying step.

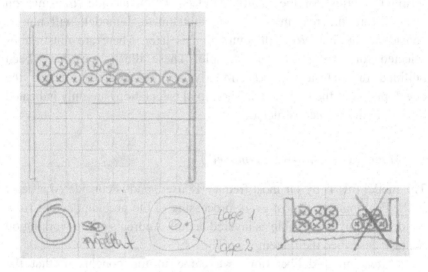

Figure 3. Real modelling of model M1

However, it was noticed that some of the students initially considered this 'real model' to be the real world, i.e. the annuli did not subjectively represent a simplification until the teacher asked "whether the purchaser would agree to circular wound cable", "how is electricity supposed to flow through such a cable" or "whether the cable is actually wound on the cable drum in such a way".

The layer width is 4.4 m and the cable diameter is 30 cm. Arithmetically, 14.6 circles fit next to each other. In most cases the students realise on their own that it would not make sense to round the value up. It is therefore possible to arrange 14 cables next to each other in each layer and 20 centimetres 'space' remains.

The difference between flange and hub radius is $R_1 - R_2 = 1.9$ m and the cable diameter shows that 6.33 layers fit on top of each other – thus, 6 entire layers if the cable does not protrude from the flange radius.

This calculation is partially transferred directly into the sketch as the following figure (Figure 4) exemplarily shows.

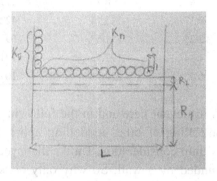

Figure 4. A sketch that summarises the modelling

The length of the entire cable is then successively calculated via the length of one single circle, the cable length in one layer and the summation of the individual layers. To determine the length of the circles in the i-th layer, the appropriate radiuses r_i are required. These are mostly calculated one after another for each layer (even in the schools for higher achieving students). The first layer is:
$r_1 = R_2 + r = 0.75$ m, thus $r_2 = r_1 + 0.30$ m $= 1.05$ m etc. Consequently, one annulus is $(2 \times \pi \times r_1) \approx 4.712$ m and the entire first layer is
$K_1 = 14 \cdot (2 \times \pi \times r_1) \approx 65.97$ m. Analogously, the second layer is 92.36 m, the third 118.75 m, the fourth 145.14 m, the fifth 171.53 m and the sixth layer is 197.92 m. The total length of the cable is $K \approx 792$ m.

Most students are not satisfied with this argumentation. Andreas's group working on the task remarked that:

"We know that 792m is just <u>one</u> value because the cable may be furled onto the roll in different ways. Besides, this is merely a rounded sum."

The following table – which has proven to be a helpful indicator for arithmetic errors as well – lists the individual steps.

Table 1

Calculations concerning M1 (each in metres and rounded to the nearest centimetres)

layer	radius	circle length	Length of layer	sum of cable length
1	0.75	4.71	65.97	65.97
2	1.05	6.60	92.36	158.34
3	1.35	8.48	118.75	277.09
4	1.65	10.37	145.14	422.23
5	1.95	12.25	171.53	593.76
6	2.25	14.14	197.92	**791.68**
7	2.55	16.02	224.31	1015.99

The 7th layer is irrelevant here and in the following models, but it is relevant to the controlling of other modelling approaches, which take 'overlapping layers' into consideration. Unlike the layer width, the space upwards is not bounded by a wall. So why only consider 6 layers? A member of Miriam's group working on the task said that

"It would be possible to reel up a 7th layer. In this case, there would be cable protruding and it might be damaged while rolling the drum."

But there are also working groups that are apparently indifferent to this problem and therefore calculate for 7 layers.

2.2 Model M1-I ("Many-circle-model" - inner radiuses)

Almost 10% of the modelling approaches, which the first author collected in a case study with 200 German students of ages 15-16, were done using the inner radius for the calculations instead of the middle circle radius, meaning $r_1 = R_2 = 0.60m$, $r_2 = r_1 + 2r = 0.90m$ etc. The students almost always associate this with a naïve (mis-) conception of the winding characteristics of a cable, since this would mean in reality that the cable is solely tenso-elastic but not impulse-elastic. It would 'evenly' abut on the hub and would merely be deformed towards the outside, not exactly advantageous for considering cables with conducting copper core but partially a quite adequate model for pipes with an accordion-like composition. The cable length that is evaluated in the model calculation (shown in Table 2) is definitely shorter and 712m.

Table 2

Calculations concerning M1-I

layer	radius	circle length	length of layer	sum of cable length
1	0.6	3.77	52.78	52.78
2	0.9	5.65	79.17	131.95
3	1.2	7.54	105.56	237.50
4	1.5	9.42	131.95	369.45
5	1.8	11.31	158.33	527.79
6	2.1	13.19	184.73	**712.51**
7	2.4	15.08	211.12	923.63

2.3 Model M1-A ("Many-circle-model" - outside radiuses)

This model, which appears with a frequency of 5% in the already mentioned case study uses the outside radiuses to calculate the circle lengths, i.e. $r_1 = R_2 + 2r = 0.90m$, $r_2 = r_1 + 2r = 1.2m$ etc., the cable is merely compressed and the cable length is analogously calculated to be larger.

Table 3

Calculations concerning M1-A

layer	radius	circle length	length of layer	sum of cable length
1	0.9	5.65	79.17	79.17
2	1.2	7.54	105.56	184.73
3	1.5	9.42	131.95	316.67
4	1.8	11.31	158.34	475.01
5	2.1	13.19	184.73	659.73
6	2.4	15.08	211.11	**870.85**
7	2.7	16.96	237.50	1108.35

2.4 Model 2 M1-H ("Helix-Model")

Concerning the circle-model, the students mostly note that the cable actually depicts the form of a helix (see Figure 5). In this case, a member of Isabel's group remarked that:

"The result is not precise because the cable is wound obliquely."

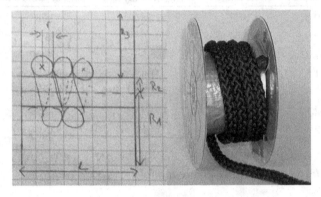

Figure 5. Cable winding as helix

The figure shows that some students speculate, partially even very complex, about the actual winding of the cable. To some extent they even discover that the 20 cm of "space" are required to actually wind the cable 14 times.

The next questions arise: Does more or less cable fit on the drum thereby? And how big is the difference? A quantitative calculation or estimation of this difference is mostly not accomplished of course because it initially requires a fundamental idea: In order to calculate the (arc) length of a helix, it is necessary to envision the lateral surface of the hub unrolled and subsequently use the theorem of Pythagoras. Thus, at this point, in time, a hint from the teacher may be necessary.

Figure 6. Calculation of the cable length representing a helix

It has proven successful to cut open a kitchen paper roll cardboard core in order to visualise the connection between the helix in space and the right-angled triangles in plane to the students (see Figure 6). The length of the hypotenuse equals the cable length which is sort after. The shorter side adjacent to the right angle is always the length 2r = 30 cm. The longer side adjacent to the right angle is the length of an annulus in the i-th layer – therefore it lies between 4.71m and 14.14m.

The calculation itself is suitable for homework. We are for now content with an estimation of the discrepancy that arises in comparison to the simple circle-model. The utmost relative difference between the longer side and the diagonal of a rectangle is shown in the first layer: $x^2_{Helix} = 4.71^2 \, m^2 + 0.30^2 \, m^2$, thus $x_{Helix} = 4.72$ m. The deviation per annulus is less than one centimetre and therefore throughout the whole first layer about 10 centimetres. The relative error is hence less than 2 per mill.

The consequences of this simplification are always overestimated. Students are amazed when they see this value, their estimations reach up to 100 metres difference. Initially, a calculation error is assumed and subsequently the derivation is surveyed once more. Finally, they argue that the difference is larger in the outer windings, because the annuli are bigger there. This is indeed true and the outmost 6th layer accounts for about 25 percent of the total length, but then again the relative error decreases to about 0.2 per mill in this case and with a value of 5 centimetres, this is less than in the first layer. The difference between the helix and the circle model is altogether less than 50 centimetres – the cable length therefore remains at about 792 m. It is actually immaterial regarding the cable length whether the cable is considered to display annuli or as a helix. The intended "aim" of the modelling via the model originator becomes apparent here: The model of "annuli" is adequate to reckon the cable **length**. It would, however, be entirely unfeasible for modelling the conductivity.

2.5 *Model M2 ("gap-model")*

This kind of modelling provides a significant attraction for this modelling example, although or rather because this approach proves to be false (see paragraph about modelling errors). For example, Miriam's group argued on the basis of the circle-model M1:

Objection: it is possible to save space if the cable is arranged in the 2nd way

Figure 7. The gap-idea

The majority presume that more cable fits on the drum by a 'clever arrangement' of the cable "wrapped around crossly". Only a few of them, mathematically higher achieving students, make the effort to calculate this model.

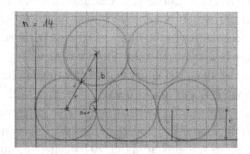

Figure 8. Calculation of the radii

The centres of the cables form an equilateral triangle (as shown in Figure 8). The following layer lies therefore not at $2r = 30 cm$ but only at $\sqrt{3}\, r \approx 1.7320\, r \approx 26 cm$ higher – thus $r_1 = 0.75 cm$, $r_2 = 1.01$ etc. Although this effectuates that less cable fits in potentially every second layer considered individually, but now 7.31 layers fit on the drum instead of 6.33 and allow one additional layer. If the mathematical model is developed, a cable length of 942 m emerges, about 150 metres more than the simple circle- or helix-model.

Table 4
Calculations concerning M2

layer	radius	circle length	length of layer	cable length
1	0.75	4.71	65.97	65.97
2	1.0098	6.34	88.83	154.80
3	1.2696	7.98	111.68	266.48
4	1.5294	9.61	134.54	401.02
5	1.7892	11.24	157.39	558.41
6	2.049	12.87	180.24	738.65
7	2.3088	14.51	203.10	**941.75**

2.6 Model MV ("volume-model I")

The following model clearly differs from the previous ones concerning the principle of the approach used. The calculations proceed from the available volume instead of following the cable routing, thus $\pi \times L \times (R_1^2 - R_2^2) \approx 81.4 \ m^3$. This is assessed in relation to the cable's cross section area $\pi \times r^2 \approx 0.071 \ m^2$. This kind of modelling usually occurs in the circle-model and helix-model namely in trying to avoid the calculations which they require.

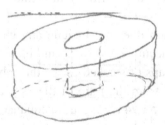

Figure 9. Real model of MV

This model is chosen with a frequency of 13% and is therefore – after the circle-model – most commonly used. The simplest kind of a 'volume-model' emanates from the idea that the cable lies on the drum without a gap. In that, the result of $81.4 \ m^3 \div 0.0707 \ m^2 =$ **1151 m** definitely constitutes an upper estimation of the cable length (provided that the cable does not protrude from the flange radius).

It is certainly not expressed in such an elaborated way when the students articulate the idea. Sometimes it partly misses the final logical consequence as the following utterance of a student shows:

"Andreas had the idea to create a volume from the drum and a piece of cable as well. We thus obtain the amount of metres which at the most fits on the roll."

However in this student report this central modelling idea can be seen. Isabel's solution (see Figure 10) follows the same route as well:

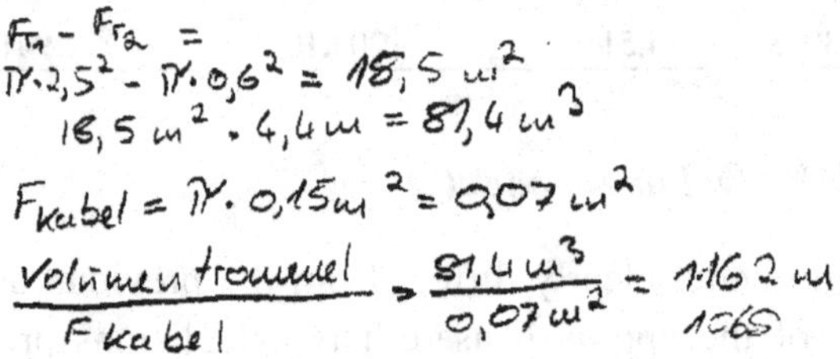

$$F_{Tr_1} - F_{Tr_2} = \pi \cdot 2{,}5^2 - \pi \cdot 0{,}6^2 = 18{,}5\ m^2$$
$$18{,}5\ m^2 \cdot 4{,}4\ m = 81{,}4\ m^3$$
$$F_{Kabel} = \pi \cdot 0{,}15 m^2 = 0{,}07\ m^2$$
$$\frac{Volumen\ trommel}{F\ Kabel} \Rightarrow \frac{81{,}4\ m^3}{0{,}07\ m^2} = 1162\ m$$

Figure 10. Estimations of Isabel

2.7 *Model MV-2 ("volume-model II")*

The students often go a step further and refine this estimation. Isabel's group continues arguing:

"Isabel's calculation does not include any gaps between the cables. The solution is inaccurate and we would like to calculate the gaps."

Considering the gaps of stacked cable, the cable length of 1151 m × 0.785 = **904 m** results from the ratio of the "square area A to the circle area A*" (thus A*/A = $\pi\, r^2/4\, r^2$) and is an enhanced upper estimation.

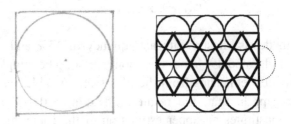

Figure 11. Estimations concerning MV-2 and MV-L

2.8 Model MV-3 ("volume-model III")

If you consider in addition to MV-2 that only 14 cables fit next to each other and 6 layers fit on top of each other inside the drum, the available volume decreases to $71.25 m^2$. The result of **791.68** m equals the exact same cable length as in model M1.

The models MV-2 and MV-3 are used significantly less frequent than the original volume model. The students are often satisfied with a 'qualitative estimation' that at least clarifies the idea of estimation of the volume model MV.

2.9 Model MV-L ("volume-gap-model")

The following modelling approaches appear even less frequently. For example one group of students – the group headed by Lars – totally disagreed with the refined volume model MV-2:

"Our group believes that more than 904 m of cable fit on it because the cable needs not to be placed by the small boxes but may also be placed in the gaps."

By transferring the idea of the cable being "wrapped around crossly" to the volume models, a picture emerges as the one displayed in right-hand part of figure 10, which is the same as model M2, having 7 layers and 14 cables in each layer. The "overlaying" semi-circle may be integrated into the rectangle in the even-numbered layers which clearly simplifies the following considerations. The actual available volume has a width of 14 cross section areas of cable and a height of 2 entire radiuses (above and below) as well as 6 reduced distances of $\sqrt{3} \times r$, which yields a volume of 75.02 m³. Equal considerations as in model MV-2 with the factor 0.785 are suitable for the top and the bottom semi-layer. The "air holes" decrease in the middle layers to the proportion between the equilateral triangle and a semi-circle (3 sectors with radius r and 60° respectively) and yield a ratio of 0.90690. Altogether, this modelling approach leads to the same length of **941.75 m** as in model M2.

2.10 *Frequency of the modelling approaches*

The modelling example was carried out with 200 students from lower secondary level from all proficiency levels. They worked in groups. Their modelling approaches were observed and, if possible, written notes collected. On the basis of classroom observations and the written approaches an evaluation concerning the developed modelling approach was carried out. The data are given in percent and rounded to half a percent. The models were only counted if they were presented as the "final" result; M1 for instance was almost always used as a basis for further considerations but often not calculated completely or merely presented as an interim result. Several models of one working group were considered, if they were offered alternatively. As already mentioned model M1 – the many circles model – was used most often, in nearly 30% of the cases.

The following diagram (Figure 12) shows the frequency scale of the different models. Typical errors are additionally listed. The data concerning the identified errors are given in percent as well. If several errors occurred in one modelling approach, they were counted repeatedly.

3 Dead Ends and Modelling Errors

The modelling example of a cable drum is especially rich, because it easily leads to dead ends. If one looks at the various modelling approaches developed, one problem can be easily detected. Is it really possible to wrap the cable using gaps as assumed in model M2 – the gap model? This model is usually convincing to everybody. In one classroom experiment the following sequence happened: An initially very cautious and thoughtfully verbalized question asked by Ines that was knowingly smiled at initially proved to be a key to dissolving the deception:

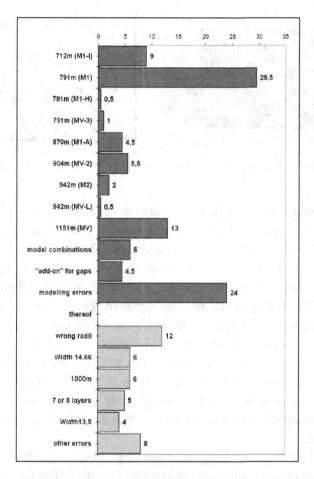

Figure 12. Relative frequencies of models and types of errors in the modelling approaches

If the first layer is wrapped to the right, so the second layer must necessarily be wrapped to the left (see Figure 13). It is not possible to use the gap since both movements are opposed to each other, a counter-rotation is necessary So, the optimal reeling strategy is to place each layer of the cable on top of each other and the 792 m of the first model is a good approximation for length.

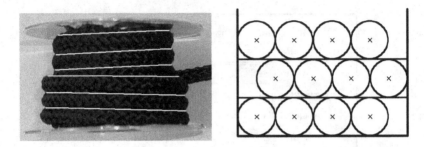

Figure 13. Counter-rotating winding and an adequate slice plane model

The modelling example of the cable drum is therefore a prototypical example of simplification of a real model without considering the conditions and restrictions of the real situation. To "wrap around crossly" is easily possible in case of the slice plane sketch, but not practicable in reality because of the necessary counter rotating windings (see Figure 13). The graphic on the right in Figure 13 paradoxically depicts an excellent model of the slice plane of the cable drum, although shifted, but always featuring 30 cm of distance because the cable of the following layer has to be placed on top of the cable lying beneath it. Following the positioning of the cable, it is possible to detect that the "unused" width is required to lead the cable into the next layer and to prevent unwanted adhesion processes, when the cable is getting warm in the summer sun.

Based on the case study the following student modelling errors on the basis of their calculated results can be identified. The following table (Table 5) summarises wrong lengths of cable that occur (more) frequently as well as their respective modelling. The abbreviation *EM-M* stands for *erroneous modelling based on model M*.

Table 5

Typical modelling errors

Cable length	Description of approach and errors made
554 m	*EM-M1*: First radius is 0.675 m = 0.6 m + 0.075 m = $R_2 + r/2$ (diameter of the hub plus half the cable radius) and further radiuses in a distance of 0.15 m (cable radius)
583 m	*EM*: width of 14.66 cables lying next to each other and a height of 3.80 m = 2 · (R1 − R2) calculated twice ("upper and lower part"), then 3.80 m : 0.3 m = 12.66 m (!) and 14.66 × 12.66 × π = 583 m
594 m	*EM-M1*: 6 layers each with a distance of radiuses of 0.15 m
692 m	*EM-M1*: same as 583 m, but including 7 layers
746 m	*EM-M1-I*: using the inside radius and calculated with 14.66 windings breadthways
763 m	*EM-M1*: calculated by the use of 13.5 windings
782 m	*EM-M1*: 2.15 m instead of 2.25 m as radius of the 6th layer
791 m	*EM-M1-I*: 8 layers (using $R_1 : 2r$); inside radius, but each with a distance of 0.15 m – "coincidentally" gives the same result as M1
923 m	Combination of M1-I and "M2": placing without gaps, thus a 7th layer fits on the cable drum but calculated by using the old radiuses (thus equals M1-I containing 7 layers)
925 m	*EM-"M2"*: alternately calculating with 14 and 13 cables per layer; altogether 95 windings; Then the proceedings become opaque: cable length = 95 × π × 3.1 m (equals $R_1 + R_2$)
971 m	Combination of MV and M1: average value of 791 m and 1151 m
996 m (1000 m)	M1 plus further considerations – e.g. that due to the gaps 29 additional cable windings of "middle length" fit on the drum; "29, therefore result 1000 m"
1000 m	Average value of MV und MV-2 resulting in 1028m and then "preventively a little less to get at 1000"
1000 m	*EM-"M2"*: Initially calculated M1 using 14.66 and (without adjusting the radiuses) added an estimated seventh winding; then rounding the 963 m to the "customary standard" of 1000 m
1094 m	MV plus "5% for the gaps"
1108 m	*EM-M1-I*: inner radiuses, 6 layers each with a distance of 60 cm
1163 m	MV1 including major rounding errors
1582 m	*EM-M1*: calculation of the perimeter using the inner and the outer radius $2\pi (r_1 + r_2)$

The choice of the radii, in particular, as well as the handling of the apparently redundant space in the layer width and flange height as well as the desire for even results cause many different "outcomes".

4 Possible Generalisations: From the Individual Case to the Formula

The calculations have to be redone from the onset if a cable drum with different dimensions appears. This may give reason to continue working with the problem in mathematically interested classes. A regularity appears while observing the radiuses:
$r_1 = R_2 + r$, $r_2 = R_2 + 3r$, $r_3 = R_2 + 5r$ etc.,
in general $r_i = R_2 + (2i-1) r$ for $i = 1, ..., n$.

Considering a cable drum containing n layers, then the cable length is:

$K \approx 2\pi m \cdot (R_2+r) + 2\pi m (R_2+3r) + 2\pi m (R_2+5r) + ... + 2\pi m (R_2+(2n-1)r)$
$\quad = 2\pi m [(R_2+r) + (R_2+3r) + (R_2+5r) + ... + (R_2+(2n-1)r)]$
$\quad = 2\pi m \cdot [n R_2 + r (1+3+5+ ... + (2n-1))]$
$\quad = 2\pi m [n R_2 + r \cdot n^2]$ (cf. Figure 14)
$\quad = 2\pi m n (R_2 + r \cdot n)$

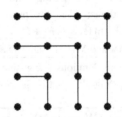

Figure 14. Sum of odd numbers

The result for the cable drum shown at the beginning is once again $K \approx 2\pi \times 14 \times 6 \times (0.6 + 0.15 \times 6) \approx 792$ m, but this time with considerably less calculation effort.

Presumably nobody reveals this unaided, at least not in 10th grade classes. However, the students occasionally notice an aspect that eases the calculation of other cable drums and provides an additional interpretation of the formula $K \approx 2\pi m n (R_2 + r n)$: They detect that the first layer along with the 6th layer have exactly the same share in the cable length as the 2nd along with the 5th and the 3rd and 4th layer respectively. They generally calculate the mean value of two layers (here: 131.95 m) and multiply this value by the amount of layers (here: 6×131.95 m = 791.70 m).

The question arises, why does the average value actually always remain constant and how could this be explained. Mathematically this is because the length of each single annuli is proportional to the radiuses. It is by no means self-evident to the students that the sum of two circumferences equals twice the circumference of the respective 'mean circle' (shown in Figure 15).

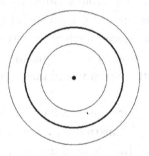

Figure 15. The circle averaging

It is therefore possible to calculate the entire cable length using a mean radius. The very same formula as above arises by using $K \approx 2\pi \times m \times n \times \bar{r}$ and a mean radius of $\bar{r} = R_2 + r n$.

In connection to this, it may even be successful to convince the students of calculating with variables which is usually not very popular: The usage of variables on the one hand produces a simple formula via conversion – this economises the calculations –, on the other hand the simplified calculation method is justified at the same time.

5 Teaching Proposals and Inner Differentiation

Based on ample experiences with the usage of this example in mathematical classrooms, one conceivable option to arrange a teaching sequence lasting for four lessons could be:
1. lesson: Hand out of the work sheets and work in small groups
2. lesson: Presentation of the results (stacked circles, estimation(s) , ...)
3. lesson: calculation considering a staggered winding, dissolution
4. lesson: enrichments and transition to possible subsequent problems.

If the aim is only to develop the simple circle-model, a reasonable result can already be achieved within two lessons. But the example opens further questions, so that an extension is possible as far as the motivation of the students supports this.

If many different solutions appear during the group work, it may be reasonable to reorganise the groups prior to the plenary discussion, namely by assigning one person of each initial group into every newly emerging group using the jigsaw method. It may be easier to explain the results to one another instead of discussing them in front of the whole class. The final plenary discussion would then function in a summarizing and reflective way.

We rarely encounter such huge cable drums since the building of new streets is at least in industrialised countries on the decrease. The garden hose drum or a fire hose drum may not be as impressive considering their dimensions, but the problem is basically the same and this kind of cable drum is easier to transport as an object of study. Alternatively, a small hoisting drum can be constructed using everyday objects: the hub consisting of 2 old CDs and a crème tin (as shown in Figures 7 and 13).

Based on our extensive classroom experience with this example, we cannot report angriness or frustration by the students if the troublesome extensions of a model prove to be 'a dead end'. The students are quite surprised if things occasionally do not work as they expected and it is definitely advisable to take a cable drum into class in order to eliminate the last doubts.

At the end we want to reflect why this kind of task is favourable for classroom usage and why almost everybody does benefit from it. It is essential that the different approaches and/or different solution processes allow a heterogeneous depth of work, although the "circle" as mathematical model is given in principle. That means the modelling can indeed be exercised at different levels: academically weak students can also detect an approach that is reasonable and convenient with the mathematical topic and efficient students can "run riot". The students thus experience *being competent* and (at least a certain) a*utonomy* during the modelling process.

Another important aspect is that the modelling is activated by one *single* question for the entire group. The differentiation of models results from the different approaches rather than from varying tasks that are differentiated according to varying abilities. Wittmann and Müller (2004) developed the concept of *'natural differentiation'* (Wittmann & Müller, 2004, p.15) as a construct for this phenomenon, since this kind of differentiation is self-evident within "natural learning" processes outside school.

Dealing with the cable drum is actually a *repetition lesson* as a start, i.e. at the end of the lesson each student repetitively practised the calculation of circles using a motivating example. The comprehensive mathematical aspects as well as modelling aspects are open to those who do not require intensive practice.

Fortunately, no trouble arises if the students calculate by wrapping the cable around crossly given that they have had to deal with the model for a long time to determine this. Actually, almost everybody falls for this trap (at first) – even established engineers and mathematicians – as the following figure (Figure 16) exemplifies.

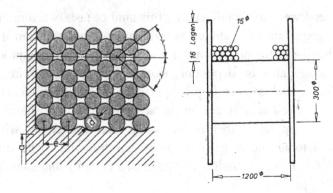

Figure 16. The "gap" occurs ...

The second graphic in Figure 16 belongs to a "closed version" of the cable drum, which might read as follows:

> "Task: 16 layers of steel wire rope with a diameter of 16 mm are wound onto a cable drum that has an internal diameter of 300 mm. The width of the drum is 1200 mm. The amount of windings in each layer differs by 1 at any one time (see sketch). Each layer lies approximately about the wire diameter "higher" than the previous one. In order to calculate the wire length, use the layer of the mean fibre."
> (Brauch, Dreyer, Haacke (1971): Mathematik für Ingenieure. Stuttgart: Teubner, p. 47, cited by Meyer, 1980, p. 86)

Although this task already specifies many modelling steps, it is not necessarily inferior to the open question of the cable drum, but follows different inner mathematical aims – in this case: the application of the arithmetic series.

We use this kind of mathematically comprehensive but closed problems in order to provide a basis for a sequence of open problems. It may be helpful to be aware of the following points, if you want to carry out the development of open problems yourself (see also Förster & Grohmann, 2008):

- Which competencies, skills or abilities can be exercised with the aid of the primary task?
- Is it possible to solve the problem by "exercising trial"? And: are there other and mathematically challenging possibilities to solve the problem?
- Are there also variations, supplementary questions of the problem? And: is it possible to summarise these questions into *one* question that opens the scope of the task and is it possible to work on this question on different levels?

The cable drum could be opened even further, for example by also allowing an arbitrary choice of the geometry of the cable drum next to the already free choice of approach. Our experiences show that this leads to too many degrees of freedom, which cause perplexity and randomness of work. The given measures of the cable drum put the mental activity into concrete terms and target the modelling approach, which is especially important for weaker students.

References

Förster, F. (2008). Die Kabeltrommel »revisited«. In A. Eichler, & F. Förster (Eds.), *Die Kompetenz Modellieren: Konkret oder kürzer* (pp. 63-68). ISTRON – Materialien für einen realitätsbezogenen Mathematikunterricht – Vol. 12. Hildesheim: Franzbecker.

Förster, F., & Grohmann, W. (2008). Möglichkeiten der Begabtenförderung im Unterricht durch natürliche Differenzierung. In M. Fuchs, & F. Käpnick (Eds.), *Mathematisch begabte Kinder* (pp. 113-123). Berlin: Lit Verlag.

Förster, F. & Herget, W. (2002). Die Kabeltrommel – Glatt gewickelt, gut entwickelt. *mathematik lehren*, issue 113, August 2002, 48-52.

Meyer, K. (1980). *Anwendungsaufgaben im Mathematikunterricht. Vol. 1 Algebra und Geometrie*. Frankfurt: Hirschgraben.

Wittmann, E. Ch., & Müller, G. N. (2004). *Das Zahlenbuch 1*. Teacher's book. Leipzig: Ernst Klett.

Chapter 16

Implementing Applications and Modelling in Secondary School: Issues for Teaching and Learning

Gloria STILLMAN

One method of engaging secondary students in mathematics classes is through taking a real world modelling approach to the teaching of mathematics, beginning in the lower years (Years 7/8/9) and developing students' modelling abilities into the senior secondary years. Of interest in this process is how teachers craft lessons and manage task sequences over time, in order to facilitate students' progress. This chapter addresses some of the issues teachers may face as they seek to implement applications and modelling within the Singapore Syllabus framework. The purpose is to reflect on approaches used in Australia and to see what lessons can be learnt from these experiences for the Singapore context. Both the teacher moves and the students' moves in this process will be examined. Specifically, the focus is on the nature of modelling and modelling tasks and how these differ from other types of tasks commonly used in secondary schools. Conditions identified as enabling the successful implementation of modelling in secondary classrooms will be discussed. Finally issues related to the assessment of student work will be addressed.

1 Introduction

Mathematical modelling connects from the outside world into the classroom. "Context-based mathematical modeling provides ideal settings to blend content and process so as to produce flexible mathematical competence." By engaging in the iterative process of modelling and its associated processes *"students develop the cognitive connections* required to understand mathematics as a discipline" (Muller & Burkhardt, 2007, p. 269).

Teachers of mathematical modelling at university level possess a worldview recognising "the power of mathematics to describe, explain, predict and control real phenomena [and appreciate] that mathematics is indispensable as a way of knowing about the world in which they live…[For them,] everything turns into mathematics." For many tertiary students, however, "it is nearly impossible to adopt this new way of looking at the world so late in one's education" (Lamon, Parker, & Houston, 2003, p. ix). It is thus important students develop awareness of this fundamentally different way of perceiving the world whilst in primary and secondary school (Stillman & Brown, 2007) or even in early childhood (Fox, 2006).

Modelling is not an add-on to the curriculum but an essential component as Mason (2001) points out. "Exposing students to mathematical modelling is not just a form of teaching applications of mathematics, nor of illuminating the mathematics being applied. It is equipping students with the power to exercise a fundamental duty" (p. 43) as a citizen to appreciate, critique and use the models and modelling that permeate and format our modern world. Successful teaching and learning of mathematical modelling is thus about developing a different worldview as both a teacher and a student. It is certainly demanding that we do more mathematics ourselves as teachers as we engage in modelling and designing tasks for our students.

2 Why Applications and Modelling in Secondary School?

Applications and modelling need to be explicitly on the agenda in secondary mathematics classrooms if being able "to solve real-world problems with messy data and to undertake mathematical modelling" (Wong & Lee, 2009, p. 36) are realisable outcomes of the Singapore mathematics curriculum framework at the pre-tertiary level. Rather than seeing these as additions to the curriculum, "applications and modelling activities should be seen as embedded within mathematics and then together the two become an effective lens for describing and analysing the real world" (Stillman, 2007, p. 466).

The most extensive and long running implementation of applications and modelling within a senior secondary curriculum is in the Australian state of Queensland where this has been a feature of the curriculum for over 20 years. This implementation is the basis of a research investigation, *Curriculum Change in Secondary Mathematics* [CCiSM] (see Stillman & Galbraith, 2009), and data from interviews conducted for this project will be reported here. When a current Queensland senior secondary classroom teacher was asked to reflect on her recollections of her and other teachers' reactions to the change from an abstract theoretical mathematics curriculum in the early 1990s to an emphasis on applications and modelling, she recalled it as an opportunity to *"Get away from it being so incredibly dry and boring which, okay, they did it to get themselves into uni but really ... I don't think there was too much enjoyment on the kid's part."* When asked if she viewed upper secondary mathematics as a hurdle to enter university studies she replied, *"Yes, yes, the filter, the filter for university studies whether you needed maths or not. If you could do this then obviously you were going to be a good student. You should be successful at university."* (CCiSM project, 2007, Teacher Interview)

The upper secondary years are also the last chance secondary teachers have to influence their students' future. Applications and modelling tasks at this level can provide motivation for continuing into mathematical careers as the following anecdote from a practising teacher illustrates.

Here's a task ... it is about traffic lights. I suppose one of the nice things is, I have been using it a number of years, one of my students when she started Year 11, they had just started matrices because all you have to do is be able to add matrices and multiply by a scalar to do it, but she loved it. She then decided "I think I'll go and do Engineering". She has just finished her Masters in Engineering and actually she is working on the tunnel in Brisbane but it was a case of this was such an exciting thing which she decided where she wanted to go with it. (CCiSM project, 2007, Teacher Interview)

Rapid advances in technology such as graphing calculators in the latter part of the 1990s acted as a lever for change in teaching approaches to accommodate these new expectations of how mathematics would be taught in senior secondary classrooms in Queensland. As pointed out by a teacher heavily involved in the curriculum implementation throughout its first 10 years,

In some ways though, the modelling came through in a de facto way with graphics calculators. So often these in-services were being conducted and the material being used was modelling material. And so it was more the technology then that was supporting, or giving teachers some idea which way to go, how they could use the technology and develop modelling. (CCiSM project, 2007, Teacher Interview)

This brought particular benefits for inexperienced and experienced teachers alike having a profound influence on the teaching of many as this highly experienced teacher at the upper secondary level points out,

Teacher: *I was very nervous at first but I really like the approach and I think it is really important for the students.*
I: *So what are those particular positives?*
Teacher: *From my point of view I think it gives us, with the technology that we have got on hand, it gives us a chance to do mathematics that has more meaning for the students. So you can use real life data. You can't do it all of the time but you can do it more and more so that the students can see mathematics in a context and they are very good thinkers. It has given me great insight into*

how students actually think. (CCiSM project, 2007, Teacher Interview)

When modelling programs have been introduced at the lower secondary level, the reaction has also been highly favourable. In the RITEMATHS research project in the Australian state of Victoria, Year 9 students were introduced to extended modelling and applications tasks over the course of one year. Of the 25 students selected to be interviewed who had experienced such a program in their school, the majority (21 out of 25) were highly positive as the following two excerpts from interviews show.

Sam: *I also like doing things like this that gives a separate theory to reality because it helps you visualise.*
I: *So the fact that this was set in a real world context?*
Sam: *It helped me visualise what the real situation could be, how we could deal with it too.* (RITEMATHS project interview, 2006)

I: *Did it help you at all? When you were trying to figure out the maths of it, did you go back and think about the real setting of it?*
Tracey: *Yeah, you do.*
I: *And that did help?*
Tracey: *Yeah, because you can imagine doing it in real life.* (RITEMATHS project interview, 2006)

A few (3 out of 25) were more ambivalent not preferring applied real world problems over abstract ones although this student did prefer the verbal nature of such tasks.

Ned: *I prefer the fact they are worded tasks rather than just blank equations on a sheet but I don't think wherever they were set would matter to me.* (RITEMATHS project interview, 2006)

Only one student did not appreciate the teacher's efforts to give the students opportunities to apply mathematics in real world settings such as soccer, bungee jumping and orienteering.

I: *What about the fact [the teacher] is trying to get you to do it in a real world setting?*
Mei: *Nuh.*
I: *You don't see that as very real world?*
Mei: *Nuh.*
I: *And do you think it is necessary to do your maths in a real world setting?*
Mei: *Not really.* (RITEMATHS project interview, 2006)

3 Applications Versus Mathematical Modelling

Locating mathematical tasks in meaningful contexts is often claimed to be enriching for students as their mathematical experiences become connected to real life experiences (e.g., Zbiek & Connor, 2006, p. 89). In the main, however, such tasks are mathematical applications connecting classroom mathematics to the outside world (Mangelsdorf, Brazil, & Tordesillas, 2003). What then is the difference between mathematical modelling tasks and mathematical applications?

In mathematical applications the task setter starts with mathematics and reaches out to reality. A teacher designing such a task is effectively asking: Where can I use this particular piece of mathematical knowledge? This leads to tasks that illustrate the use of particular mathematics content. They are a useful bridge into modelling but are not modelling in themselves as seen in the examples below. In the first example, graphs of functions could be linked to architectural design in the roof of a railway station under construction. In the second example, a trigonometric equation is used to describe the pitch of the sound from a security alarm.

| We are studying graphs of functions. What in the real world can I use to show the utility of mathematics? |  |

Figure 1. An application of functions to the roofing structure at Southern Cross Railway Station, Melbourne

A family worried about the increase in crime in the city, decides to install a loud security alarm. The pitch, P, in hertz, of the sound made by the alarm varies with time, t minutes from the start of the alarm. The pitch of the sound can be described by the equation:

$P = -16 \cos(\pi/6)t + 2400$

The family goes on holidays for a long weekend leaving the teenage son, Chris, at home. Chris, after a long night at the clubs can hear sounds only above 1600 hertz.

Determine the greatest time a burglar has to break in, steal the DVD player and get away before Chris is aware he has been robbed. Justify your answer by drawing a sketch and providing a suitable explanation.

Figure 2. An application of a trigonometric function to a burglary situation
(Source: Teacher task from CCiSM project)

With mathematical modelling on the other hand, the task setter starts with reality and looks to mathematics before finally returning to reality to judge the usefulness and desirability of the mathematical model for description or analysis of a real situation. A teacher in designing such a task is thus asking: Where can I find some mathematics to help me with this problem? A first example comes from rural Australia where the health of some of our rivers and wetlands is now being threatened by some of the agricultural practices used in the past. Agricultural scientists have encountered a salinity problem in the wetlands affecting biodiversity and are using mathematical modelling to help solve this. The question to be asked is: How can I model this with mathematics to suggest future scenarios? The scientists did indeed construct two conceptual models to test experimentally (Department of Sustainability and Environment, 2006, 2007). The second example involves a forensic investigation for a murder trial (Cross, 2006, 2009; King, 2007) as a source of ideas for a mathematical modelling task (see Figure 3) providing a situation that in this case is accessible to upper secondary school mathematics such as projectile motion and vector calculus.

How Did the Body Get There?

Rene Rivkin's former chauffeur Gordon Wood murdered his model girl friend Caroline Byrne because she could have revealed lies he told about a secret shareholding in Offset Alpine Printing, or disclosed information that a fire at the plant was a "set-up", a court has heard.

Gordon Wood, 44, was charged with murdering a 24-year-old female model whose body was found at the bottom of a cliff at Watson's Bay, in Sydney's east, on June 8, 1995. The case was originally treated as a suicide, but after a lengthy investigation, Mr Wood was arrested in London and extradited to face the murder charge. Crown prosecutor (Mark Tedeschi, QC) said evidence compiled by University of Sydney physics professor Rodney Cross would show that Byrne could not have jumped or fallen from the 29 m cliff and landed wedged head-first in a crevice 11.8 m out from the cliff face.

'The only way she could have ended up so far from the cliff face was if she was forcefully thrown by a particularly strong person using a particular spear-throw technique' (According to Cross where a person is grabbed by the groin and neck and thrown from chest height using a shot put action – was the only way Byrne's body could have achieved the necessary speed and orientation to land so far out). Mr Tedeschi said Mr Wood possessed such strength because he was a fitness trainer who lifted weights four or five days a week.

Professor Cross said a body of 61 kg required a launch speed of 4.5 m/s to reach 11.8 m away from the face. Byrne weighed 57 kg. Dense foliage at the top of the cliff in 1995 meant only a 4 m run up, effectively preventing a jumper from reaching the necessary speed. Prof Cross said that a two-person throw or an underarm throw by one person would not reach the required launch speed.

Figure 3. How Did the Body Get There? (Source: *The Australian,* June 13, 2007)

4 The Modelling Process

A vast variety of diagrams have been used over the years to capture the essence of what happens during mathematical modelling to act as a scaffold in talking about modelling, designing tasks for modelling and implementing modelling in the classroom. Figure 4 is a version of the cycle used in Stillman, Galbraith, Brown and Edwards (2007). The entries A-G represent stages in the modelling process, where the thicker arrows signify transitions between the stages. The total solution process is described by following these arrows clockwise around the diagram from the top left. It culminates either in the report of a successful modelling outcome, or a further cycle of modelling if evaluation indicates that the solution is unsatisfactory in some way. The kinds of mental activity that individuals engage in as modellers attempt to make the transition from one modelling stage to the next are given by the broad descriptors of cognitive activity 1 to 7 in Figure 4. The light reverse arrows emphasise that the diagram is a simplification of the modelling process which is far from linear, or unidirectional. The light arrows also indicate the presence of reflective metacognitive activity.

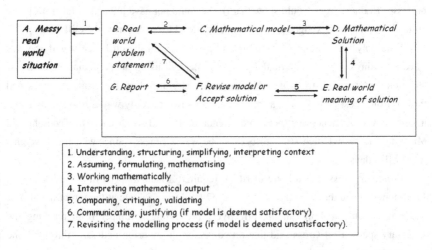

Figure 4. Modelling cycle from Stillman, Galbraith, Brown and Edwards (2007).

The transitions between the various stages become the focus for teaching. Teachers need not only engender a good sense in students of the process as a whole but also they should facilitate students' development of the processes required to make these transitions. Thus, what is of relevance is how to facilitate students' crossing the transitions during modelling. These transitions following Figure 4 are:

1. MESSY REAL WORLD SITUATION → REAL WORLD PROBLEM STATEMENT
2. REAL WORLD PROBLEM STATEMENT → MATHEMATICAL MODEL
3. MATHEMATICAL MODEL → MATHEMATICAL SOLUTION
4. MATHEMATICAL SOLUTION → REAL WORLD MEANING OF SOLUTION
5. REAL WORLD MEANING OF SOLUTION → REVISION OF MODEL OR ACCEPTANCE OF SOLUTION
6. ACCEPTANCE OF MODEL → COMMUNICATION OF A REPORT
7. REVISION OF MODEL → REAL WORLD PROBLEM STATEMENT

5 Teaching and Learning through Modelling to Increase Cognitive Demand

Modelling should not be seen as an addition to the current practice in mathematics classrooms but rather an alternative approach requiring a different worldview about what is worthwhile mathematics in secondary school and what is a sensible way to teach it. Bearing this in mind, two questions arise when changing classroom approaches to mathematics from an abstract, drill and practice based approach to one that values modelling. The teaching question of relevance is: How can teachers engineer the learning environment to enable the implementation of tasks of higher cognitive demand with applications and/or modelling playing a central role? Regarding students, the question that arises is: How can students learn to respond to modelling tasks of increasing complexity and cognitive demand?

From a teaching perspective, cognitive demand is mediated through careful tuning by the teacher of the interplay between (a) task scaffolding, (b) task complexity and (c) complexity of technology use (Stillman, Edwards, & Brown, 2004). Task scaffolding is the degree of cognitive processing support that the task provides task solvers, enabling them to solve complex tasks that would be beyond their capability if they depended on their own cognitive resources. It is meant to be scaffolding—support that is provided during the beginning stages but then there is fading of that support as students grow in their understandings. Task scaffolding comes through (a) how the task is structured, (b) the type of technology chosen, (c) whether the task is technology enabled or by-hand methods are privileged by the task, and (d) whose choice it is to decide (a) to (c) – the task solver or the task setter.

Decisions for teachers' management of cognitive demand of learning experiences are a delicate balancing act of several competing demands especially at the beginning of learning new approaches such as modelling. These competing demands include (a) maintaining higher cognitive demand versus the level of scaffolding necessary to ensure progress and engagement by all students, (b) complexity of the task versus the sophistication of the required knowledge base of the student (both technological and mathematical) and (c) the complexity of technology use necessary versus student facility with, and knowledge of, technology. How teachers fulfilled these competing demands was studied in the RITEMATHS project.

6 Conditions for Success in Modelling in Secondary School

Several conditions for successful implementation of modelling tasks into mathematics classrooms in a program introducing students to modelling over the course of their Year 9 program were identified by researchers and the teachers in a range of schools in the RITEMATHS project. The intention of the project was that the cognitive demand of these tasks would be mediated by the teachers, as necessary, depending on where the students were in the development of modelling competencies but cognitive demand would be kept at a high level particularly towards the end of the task sequence. These conditions fall into three categories: (a) the structure or nature of the task, (b) student conditions and (c) teacher conditions.

6.1 Conditions related to structure or nature of the task

a. *Allowing students to fly*. Tasks that were designed so as to enable all students to go beyond what was the anticipated level of achievement at the end of the year were seen as critical for the program to be challenging all students to increase their engagement cognitively with the tasks, albeit to different levels of depth. For the high performing students the task sequence over the year was meant

> "*to ensure ... that the technology still allows them to move on to whatever their next window of learning is as opposed to trying to keep them at a level of learning where everyone has to be*" (RITEMATHS project, 2005, Teacher Interview).

It was, however, a sequence of tasks that allowed all students opportunities to extend the ceiling of their performance and understanding in mathematics as the same teacher points out as he continues:

> *There's certainly a common view of where everyone should be at the end of Year 9 but everyone's still going to be stretched beyond this. A nice example of it I heard recently was the three window approach. You know students at the beginning [are] at the view where you want them to be and students who are looking at other views upstream of where you want them to be and that upstream next year is where you want them to be. You don't want to hold back these students who can independently creatively think at some of these open-ended tasks. That's what we're trying to use them for, not just as a learning task but also it will allow them to investigate*

deeper and deeper into their understanding of mathematics and why it is they write it.

b. *Fostering natural curiosity.* Modelling tasks that can foster students' natural curiosity to make conjectures and test these during their explorations are intrinsically engaging. Toh (2009) cites Schmidt and Lahroodi (2008) and paraphrases their definition of curiosity as "a motivationally original desire to know" adding "this desire arises and, in turn, sustains one's attention and interest to know" (p. 244). In open modelling tasks students choose the question they want to answer about the real world situation they are investigating and this is obviously contingent on their natural desire to explore their world and pursue answers to questions that are of interest to them personally. In tasks chosen by the teacher where the modelling is more structured there is the possibility that students will not have this intrinsic interest and be motivated to engage as their curiosity has not been piqued. Thus, the choice of a range of tasks with different real world situations to model should be a part of any modelling program but even then, those that have the potential to foster natural curiosity will be the better to choose.

c. *Allowing freedom in choosing technology.* Rapid changes in technology have acted as an enabler of modelling in those educational systems where modelling has been implemented system-wide such as Ontario (Suurtamm & Roulet, 2007) and Queensland (Stillman & Galbraith, 2009). However, if students are to develop as independent modellers they must be allowed freedom in choosing the technology to use in the task to enhance modelling aspects. To do this effectively, however, they need to know the different affordances of technology. A function grapher, for example, enables different analysis from a graph plotter. Both teachers and students need to be cognisant of these differences to be able to make judicious choices between technologies when modelling.

d. *Allowing use of multiple representations supporting connection making.* As Lee points out, in modelling tasks "sense-making should come first and the content second" (2009, p. v). This is particularly true if students are going to successfully bridge the various transitions in the modelling cycle pointed out above. Modellers need to be continually aware of the connections between the situation being modelled and their mathematical activity. One way to ensure this is to use tasks that allow multiple representations (e.g., graphical, algebraic, numerical) where initially the teachers scaffold the use of many representations and the drawing of connections between them. In a task about orienteering, for example, a student might produce a tabular representation of the distances a runner might have to run to travel to different drink stations in a section of the course that passes through a field. As well as this numerical representation, students could be asked to produce a scale diagram of paths through the field, and a graph of the distance travelled versus the distance of each station from the beginning of the field. Drawing connections between these various representations helps the students deepen their understanding of the situation and their sense of meaning of the situation.

e. *Requiring the answering of interpretive questions.* Interpretive questions that students are required to answer at critical junctures in the modelling process and at the end of their modelling serve two purposes. Firstly, these questions help in developing the reflective activity that accompanies the metacognitive competencies required in modelling. Secondly, they foster the drawing of conclusions based on evidence that are required in the second last transition for communication of the results from the modelling. To serve the first purpose, when beginning a program of modelling, reflective questions can be asked at strategic points to provoke students to pause and reflect on their immediate progress in relation to the total modelling purpose as shown in Figure 4. Young modellers benefit from such activities and appreciate their self-regulatory activity being scaffolded. However, they also want the freedom that comes as such scaffolding is faded across a series of tasks so they can make their own decisions about their modelling pathways. In relation to the second purpose, Krajcik and Blumenfeld (2006, p. 324), point out that "drawing and justifying conclusions using primary evidence requires sophisticated thinking and much experience", experience that needs to be provided often. Recipe approaches to modelling or problem solving do not allow students the opportunity to use and infer from real data answering their own questions. Students initially will need support in articulating explanations and drawing conclusions and that is the place of interpretive questions.

f. Scaffolding of recording of key mathematics. In an era when technology use is becoming more and more prevalent, ensuring that students record key information such as the assumptions they have made, how they came up with estimates or key working out becomes of greater importance. The mere recording of interim results or answers does not leave a trace of the mathematical thinking behind these. However, as Loh et al. (2001) noted in the context of school science, there are two challenges students face when engaging in extended investigations for the first time. These are an inability to recognise when to keep records and failure to plan and monitor progress effectively. It is thus prudent for teachers designing extended modelling tasks for students beginning a modelling program, initially at least, to provide timely instructions supporting recording of key information, a planned solution, checking and verification of results. As student task expertise grows, "fading" of this scaffolding should occur (Guzdial, 1994) otherwise it becomes an exo-skeleton holding up the whole edifice of performance on such tasks and independent modelling is not occurring.

6.2 *Student conditions*

a. *Developing understanding of situation in groups.* The formulation phase of modelling (i.e., making the first two transitions successfully) is notoriously difficult (Stillman & Brown, 2007). At lower secondary level it is important that students work collaboratively in small groups and as a class to develop understandings of the situation. These tasks often require domain knowledge about the situation in question and this is more likely to be activated in a group or class discussion situation. In a task modelling a shot at goal in soccer, for example, students were taken out onto a nearby soccer field and several students experienced and described for those watching the apparent closing down and then enlarging of the face of the goal as they ran towards the goal line on a line perpendicular to it (See Stillman & Brown, 2007).

b. *Using physical activities related to the task to develop domain knowledge.* Often modelling tasks involve domain knowledge related to the situation that is quite subtle in conveying its relevance. To help alert younger students to the subtleties in the real situation of tasks such as shooting a goal in hockey or soccer or choosing a drink station in an orienteering race, teachers engaged their students in physical activities in and outside the classroom which gave further insight into the situation. To develop understanding of how the angle subtended by the face of an opening such as a goal changed for viewers depending on where they were positioned in relation to the opening, one teacher set up a small goal in the classroom and asked students to attempt to throw a tennis ball through the opening from various positions around the room. They then were asked if, in order to win a small prize, they would throw the ball themselves (from a difficult angle) or pass to their friend (who had a better angle). When they decided to pass the ball they had to explain why and in the process articulated the effect of position on angle.

c. *Participating in rich dialogue and discussion with peers and the teacher.* Rich discussions through debate, argumentation, and spelling out of points of view for others to critique are pivotal in communicating and developing understanding and choosing strategies throughout the task as this Year 9 student points out:

> Sam: *... and checking within our group too and if we had a dispute we'd not really lecture each other but we'd, you know, not really [have] an argument but we would sort of explain all our points ... like say, I think Kia was going to use Pythagoras and I was going to use tan. He sort of explained why he used Pythagoras. I explained why I used tan. You know we worked out that it was going to be inverse tan. Yeah, so that's how we sort of checked if like we were right or not.*

6.3 Teacher conditions

a. *Knowing when to intervene.* Perception on the part of the teacher so as not to stifle student growth by intervening too soon or too often and thus not allowing students' the space to struggle is crucial for modelling tasks. However, Blum and Leiss (2007) point out that teachers often lack knowledge of strategies for such "independence-supporting" interventions in demanding mathematical tasks (p. 230).

b. *Positive expectations of student engagement with modelling.* It is also necessary that teachers convey to students that they expect they can model and they will do so. For teachers to believe that only certain students are able to model is not realistic as all students can produce a model based on the mathematics they are able to activate. These models may have various levels of sophistication but they are models nonetheless.

c. *Knowing the essence of the task.* It is imperative that teachers know the essence of the task from the perspectives of mathematics, technology (if being used) and investigative and modelling skills development. If they do not know, teachers could leave out critical elements of the task which undermines its integrity as a modelling task or tell students to do things rather than allow them to choose to do as is necessary in developing independent modelling competencies.

d. *Tolerating different rates of progress.* Confidence on the part of the teacher to be at ease in a classroom where students are progressing through a task at various rates is essential when taking a modelling approach. As soon as a teacher attempts to reign students in and control the rate of progression a damper is placed on alternative pathways and very quickly the teacher's way of approaching the task is conveyed to students and adopted by most. If the goal is seen as learning the process of modelling rather than successfully completing this task using a prescribed set of mathematical content (Lee, 2009), the need for all to progress together to the same end point dissipates. What is then a desirable outcome is that all students progress in their modelling where developing fluency in the processes involved in the transitions (see Figure 4) are viewed as an integral part of mathematics itself. The problem itself might not be solved by some or all "but through the process of modelling, a better understanding of the problem is achieved" (Ang, 2009, p. 160).

e. *Wrapping up the task.* It is important that there be a final discussion and sharing of ideas as a whole class as this is often when both teacher and student make connections to what else is happening in mathematics at the time. Leaving a task hanging incomplete, or changing direction immediately to another topic, conveys the impression that the purpose of modelling is to work on, or solve, the particular task at hand rather than this modelling experience being another progression in the students' developing of modelling competencies. The goal is that the students can model independently in the long term.

7 Assessment Issues

Assessment is a major issue when it comes to modelling as it can be a challenge to conduct and a driver of curriculum change (Galbraith, 2007; Stillman & Galbraith, 2009). Modelling processes are mathematical content in their own right and should be assessed. If this is taken on board then there is no imperative for modelling assessment tasks to incorporate testing of the latest content topic taught. Furthermore, if modelling is seen as an underpinning principle to the mathematics curriculum it should not be an optional alternative in assessment but a necessary inclusion. Modelling cannot be assessed validly and reliably in timed examination situations; rather modelling projects should be done in groups and students be allowed extended time to complete these including out of class time. To assess such tasks criteria and standards can be used as shown in Table 1.

Table 1
Standards and performance descriptors for a modelling task

Standard	Description of Performance
A	The overall quality of the students' response generally demonstrates mathematical thinking which includes: • Interpreting, clarifying and analysing, identifying assumptions, parameters and variables in a range of real world problem situations; • Articulating the effects of using various assumptions; • Selecting and using effective problem solving and modelling strategies; • Selecting and using mathematical models in a range of real world situations of differing complexity; • Articulating the strengths and limitations of models that have been given or developed; • Interpreting the results of solving mathematical models in a range of real world situations of increasing complexity; • Checking the validity of models used.
B	The overall quality of the students' response generally demonstrates mathematical thinking which includes: • Interpreting, clarifying and analysing, identifying assumptions, parameters and variables in routine real world situations and in simple, non-routine real world situations; • Selecting and using effective problem solving and modelling strategies; • Selecting and/or using mathematical models in a range of real world situations of differing complexity; • Interpreting the results of solving mathematical models in a range of real world situations of increasing complexity;
C	The overall quality of the students' response generally demonstrates mathematical thinking which includes: • Interpreting, clarifying and analysing simple, routine real world situations; • Selecting and/or using mathematical models in simple, routine real world situations;

	• Interpreting the results of mathematical activities in simple, routine real world situations.
D	The overall quality of the students' response sometimes demonstrates mathematical thinking which includes: Use of given mathematical models and following of simple problem solving strategies in real world applications
E	The overall quality of the students' response indicates the student implements simple mathematical procedures but is unable to develop basic problem solving strategies.

8 Concluding Remarks

Mathematical modelling is a new and exciting field that has recently been added to the mathematics curriculum framework in Singapore. To teach effectively from a modelling perspective requires the adoption of a different worldview with respect to mathematics than teachers and students who see mathematics as essentially abstract are used to. For teachers to make the transition effectively requires them to experience modelling for themselves and develop their own skills in mathematical inquiry and modelling over an extended period of time during which they share experiences with colleagues and develop solutions to problems they pose about real world situations.

Acknowledgement

The RITEMATHS research project was supported under the Australian Research Council's Linkage-Projects funding scheme (project LP0453701). Views expressed here are the author's and are not necessarily those of the Australian Research Council.

References

Ang, K.C. (2009). Mathematical modelling and real life problem solving. In B. Kaur, B.H. Yeap, & M. Kapur (Eds.), *Mathematical problem solving* (pp. 159-182). Singapore: World Scientific.

Blum, W., & Leiss, D. (2007). How do students and teachers deal with modelling problems? In C. Haines, P. Galbraith, W. Blum, & S. Khan (Eds.), *Mathematical modelling (ICTMA 12): Education, engineering and economics* (pp. 222-231). Chichester, UK: Horwood.

Cross, R. (2006). Fatal falls from a height: Two cases. *Journal of Forensic Science*, $51(1)$, 93-99.

Cross, R. (2009). *Evidence for murder: How physics convicted a killer*. Sydney: New South.

Department of Sustainability and Environment. (August, 2006). *Wetlands, boiodiversity and salt*. Melbourne: State of Victoria. Available from www.dse.vic.gov.au/ari/

Department of Sustainability and Environment. (July, 2007). *Wetlands, biodiversity and salt: Responses to salination – an experimental approach*. Melbourne: State of Victoria. Available from www.dse.vic.gov.au/ari/

Fox, J. (2006). A justification for mathematical modelling experiences in the preparatory classroom. In P. Grootenboer, R. Zevenbergen & M. Chinnappan (Eds.), *Identities, cultures and learning spaces*, Proceedings of the 29th annual conference of the Mathematics Education Research Group of Australasia, Canberra, (Vol. 1, pp. 221-228). Adelaide: MERGA.

Galbratih, P. (2007). Assessment and evaluation – Overview. In W. Blum, P. Galbraith, H-W. Henn, & M. Niss (Eds.), *Modelling and applications in mathematics education: The 14th ICMI study* (pp. 405-408). New York: Springer.

Guzdial, M. (1994). Software-realised scaffolding to facilitate programming for science learning. *Interactive Learning Environments*, 4, 1-44.

King, D. (2007). Wood 'killed Byrne to hide his lies'. *The Australian*, June 13. Available from www.theaustralian.news.com.au/

Krajcik, J. S., & Blumenfeld, P. C. (2006). Project-based learning. In R. K. Sawyer (Ed.), *The Cambridge book of the learning sciences* (pp. 317-333). New York: Cambridge University Press.

Lamon, S., Parker, W., & Houston, K. (Eds.) (2003). *Mathematical modelling: A way of life*. Chichester, UK: Horwood.

Lee, P. Y. (2009). Foreword. In K.C. Ang (Ed.), *Mathematical modelling in the secondary and junior college classroom*. Singapore: Prentice Hall.

Loh, B., Reiser, B., Radinsky, J., Edelson, D., Gomez, L., & Marshall, S. (2001). Developing reflective inquiry practices: A case study of software, the teacher, and students. In K. Crowley, C. Schunn, & T. Okada (Eds.), *Designing for science: Implications from everyday, classroom, and professional settings* (pp. 279-323). Mahwah, NJ: Erlbaum.

Mangelsdorf, C., Brazil, M., & Tordesillas, A. (2003). Getting hooked on mathematics through its applications. *Teaching Mathematics and Its Applications, 22*(2), 63-75.

Mason, J. (2001). Modelling modelling: Where is the centre of gravity of-for-when teaching modelling? In J. Matos, W. Blum, K. Houston & S. Carreira (Eds.), *Modelling and mathematics education* (pp. 39-61). Chichester, UK: Horwood.

Muller, E., & Burkhardt, H. (2006). Applications and modeling for mathematics — Overview. In W. Blum, P.L. Galbraith, H-W. Henn, & M. Niss (Eds.), *Modelling and applications in mathematics education: The 14th ICMI study* (pp. 267-274). New York: Springer.

Stillman, G. (2007). Upper secondary perspectives on applications and modelling. In W. Blum, P. L. Galbraith, H-W. Henn, & M. Niss (Eds.), *Modelling and applications in mathematics education: The 14th ICMI study* (pp. 463-468). New York: Springer.

Stillman, G., & Brown, J. (2007). Challenges in formulating an extended modelling task at Year 9. In H. Reeves, K. Milton, & T. Spencer (Eds.), *Mathematics: Essential for learning, essential for life* (pp. 224-231). Adelaide: AAMT.

Stillman, G., & Galbraith, P. (2009). Softly, softly: Curriculum change in applications and modelling in the senior secondary curriculum in Queensland. In R. Hunter, B. Bicknell, & T. Burgess (Eds.), *Crossing divides*, Proceedings of the 32nd annual conference of the Mathematics Education Research Group of Australasia, Canberra, (Vol. 2, pp. 515-522). Adelaide: MERGA.

Stillman, G., Edwards, I., & Brown, I. (2004). Mediating the cognitive demand of lessons in real-world settings. In. B. Tadich, S. Tobias, C. Brew, B. Beatty, & P. *Sullivan (Eds.), Towards excellence in mathematics,* Proceedings of the 41st Annual Conference of the Mathematical Association of Victoria (pp. 487-500). Melbourne: MAV.

Stillman, G., Galbraith, P., Brown, J., & Edwards, I. (2007). A framework for success in implementing mathematical modelling in the secondary classroom. In J. Watson & K. Beswick (Eds.), *Proceedings of the 30th annual conference of the Mathematics Research Group of Australasia (MERGA)* (Vol. 2 pp. 688-707). Adelaide: MERGA.

Suurtamm, C., & Roulet, G. (2007). Modelling in Ontario: Success in moving along the continuum. In W. Blum, P. Galbraith, H-W. Henn, & M. Niss (Eds.), *Modelling and applications in mathematics education: The 14th ICMI study* (pp. 491-496). New York: Springer.

Toh, T. L. (2009). Arousing students' curiosity and mathematical problem solving. In B. Kaur, B.H. Yeap, & M. Kapur (Eds.), *Mathematical problem solving* (pp. 241-262). Singapore: World Scientific.

Wong, K. Y., & Lee, N. H. (2009). Singapore education and mathematics curriculum. In K. Y. Wong, P. Y. Lee, B. Kaur, P. Y. Foong, & S. F. Ng (Eds.), *Mathematics education: The Singapore journey* (pp. 13-47). Singapore: World Scientific.

Zbiek, R., & Connor, (2006). Beyond motivation: Exploring mathematical modeling as a context for deepening students' understandings of curricular mathematics. *Educational Studies in Mathematics, 63*(1), 89 -112.

Part IV

Conclusion

Chapter 17

Mathematical Applications and Modelling: Concluding Comments

Jaguthsing DINDYAL   Berinderjeet KAUR

In this chapter, we discuss some questions that we feel many stakeholders would ask, such as: Why include applications and modelling in school mathematics? How can applications and modelling be brought into the mathematics classroom? We briefly review some issues about the implementation of applications and modelling in the local curriculum. We also speculate the future of applications and modelling in the Singapore school mathematics curriculum.

1 Introduction

The different chapters in this book amply demonstrate that applications and modelling have an important role in the teaching and learning of mathematics. The authors of the chapters have carefully highlighted the ways in which applications and modelling can be used at both the primary and secondary levels. There are insightful comments and suggestions about classroom practice relevant for the school mathematics teacher. There are also abundant ideas for further research in classrooms. However, the issue of applications and modelling in the school curriculum is far from being resolved. Blum, Galbraith, Henn and Niss (2007) have aptly cautioned that:

Yet while applications and modelling play more important roles in many countries' classrooms than in the past, there still exists a substantial gap between the ideas expressed in educational debate and innovative curricula on the one hand, and everyday teaching practice on the other. In particular, genuine modelling activities are still rare in mathematics classroom. (p. xi)

The question to ask then is: Where are we headed with applications and modelling in the school mathematics curriculum? There are several issues associated with the terms *applications and modelling*, although at first glance these terms may seem to be familiar to everybody. Almost always, questions asked about applications include amongst others: What are applications? Why include applications in school mathematics? How can applications be brought to the classroom? What issues are related to applications? (see Sharron & Reys, 1979, p. vii). Inevitably, one may also ask: What is modelling? Why include modelling in school mathematics? How can modelling be brought to the classroom? What issues are related to modelling? In addition, one may also ask: How do we assess students' performance in these domains? We need answers to these legitimate questions about applications and modelling if we wish to foresee a successful implementation in schools. In what follows, we attempt to answer these questions about applications and modelling – termed *topics*, for easier reference in text, although they are essentially processes that cut across content domains.

2 Applications

Niss, Blum and Galbraith (2007) claimed that an application of mathematics occurs every time mathematics is applied, for some purpose, to deal with some domain of the extra-mathematical world. Thus, the term application involves focusing on the direction of "mathematics to reality". From this perspective, most of the non-routine problems used in the problem-solving curriculum in Singapore will fall in this category of application of mathematics. **Stillman** in her chapter in this book, pages 300 - 322, uses a similar description for application.

2.1 Why include applications in school mathematics?

Historically much of mathematical knowledge evolved from mankind's quest for a better life. Although mathematics has empirical origins, the polished presentations in school textbooks portray mathematics as an abstract subject detached from reality. Perhaps the most important reason for including applications in school mathematics is that we wish to demonstrate to students that, after all, mathematics is a useful subject. In doing so, the expectation is that students will find mathematics relevant to their everyday life and be more and more motivated to study the subject. Cognitively, applications are more demanding of students as they involve higher order thinking skills which is another reason for their inclusion in the curriculum.

2.2 How can applications be brought to the mathematics classroom?

Applications are not new in the mathematics curriculum. Some aspects of applications are included in mathematics curricula all over the world. However, what is meant by applications, how much of applications need to be taught and how applications have to be taught are open to debate. In Singapore, because of the strong focus on problem solving students on the average have been involved in solving some harder problems, leaning more towards the applications side. There is no doubt that the teaching of applications involves students in some form of higher order thinking skills which is highly valued in mathematics. Applications are not meant only for the more able, there are aspects of applications in mathematics that can be adapted to suit the needs of students of all abilities. The challenge for implementing applications in the classroom then rests on the shoulders of the teacher. Sharron and Reys (1979) claimed that:

> The vastness and depth of appropriate applications generate a feeling of uneasiness for many mathematics teachers. Preparing to teach mathematics makes demands on a sizeable share of one's formal education, leaving little or no time for comprehensive experiences in areas where real-life applications in mathematics abound. A lack of such experience may inhibit a teacher from veering off from the

abstract or pedantic to the practical or serviceable areas of mathematics. (p. vii)

A teacher who is himself or herself confident in the applications of mathematics and has a strong content background with a sound pedagogical content knowledge has better chances of implementing the applications of mathematics in his or her classroom. This has implications for teacher preparation. Teachers in Singapore are exposed to problem solving and applications but not quite to modelling during their pre-service programmes. For the implementation to be successful, mathematics teachers need to become more confident to teach these topics.

2.3 *How do we assess students' performance in applications of mathematics?*

One approach is certainly to assess the product of the learning process. The implication being that a good product results form a good process. However, we know that this logic is not necessarily valid. There is a perennial *process-product* debate in the learning of mathematics. Another issue is whether applications in mathematics are meant only for individual students, only for groups of students or for both individual and groups of students. The approach used for teaching individual or groups of students in class will have implications for assessment. For example, how do we assess both individual and group performance in applications tasks in mathematics? Essentially we should value both individual and group work as working collaboratively is an important 21st century skill.

3 Modelling

Alongside applications, the framework of the school mathematics curriculum (see p. 5 of this book) also emphasises modelling. Modelling, just like applications, is not a stand-alone topic in the framework. It cuts across different topics. In very simple terms we can think of modelling as starting with the identification of a real-life problem. The modeller then tries to create a model of the situation by identifying key variables and the relationships among these variables. The problem is then solved as a purely mathematical problem and the solution interpreted in the real-life context which generated it. The modeller then attempts to make sense of the model and improve the model. The obvious thing to note here is that a model or modelling is from the perspective of the modeller and is not something universal.

3.1 *Why include modelling in the mathematics curriculum?*

There is always a debate when something new is introduced into the curriculum. The newness of any topic is a challenge for implementation, as teachers are generally content doing what they have always been doing for many years. Although, modelling may seem new, elements of the modelling perspective have always existed in the mathematics curriculum. The renewed interest in modelling stems from corresponding changes taking place in mathematics curricula all over the world. Mathematics educators wish to make mathematics more meaningful to students by highlighting the importance of mathematics in solving real-world problems (see Dindyal, 2009).

3.2 *How can modelling be brought to the classroom?*

Just as for applications, the curriculum documents from the Ministry of Education (2006a, 2006b) do not provide any specific implementation guidelines for modelling. Much is left to individual interpretations of teachers or textbook writers. If modelling skills have to be developed by students then teachers who are going to implement the curriculum must themselves develop those skills. Currently, very few mathematics teachers have received any form of training for teaching modelling at primary or secondary levels. Curriculum developers must include modelling tasks for students in the textbooks which teachers can use or adapt for their classrooms.

3.3 *How do we assess students' performance in modelling tasks?*

Following from the dictum, "what gets assessed gets taught", we have to carefully look into assessment if we wish students to acquire the valued modelling skills. Certainly, some assessment guidelines need to be provided as modelling tasks cannot be assessed in the same manner in which regular mathematical problems are assessed. Assessing only the product of the learning process may be a very simplistic approach to assess modelling skills. A kind of analytical rubric that will assess students' performance at the various stages of the modelling process needs to be developed besides also assessing students' affect.

4 The Mathematics Curriculum

A curriculum is never static. It is always in a state of flux, invariably responding to the rapid changes in the economic and socio-cultural environment of which it is a part. Some changes are minor and minimally affect the school system. On the other hand, there are changes which are quite encompassing and which severely affect the school system by placing undue pressure on all stakeholders and on teachers, in particular, for implementation. The change associated with applications and modelling in the mathematics curriculum is somewhere in between these two extremes. Applications and modelling in the mathematics curriculum refer neither to minor curricular adjustments nor to major ones and yet common sense, if nothing else, tells us that they are important in the mathematics education of any child.

Global trends have always impacted the school mathematics curriculum in Singapore. The curricular reforms that resulted from the *New Math* movement in the 1960s paid very scant attention to applications and modelling. The focus for that reform was the structure of the discipline with an emphasis on new content areas in school mathematics. Towards the end of the 1970s, it was clear that applications, in particular, needed a stronger foothold in the school curriculum. As a result the National Council of Teachers of Mathematics (NCTM) devoted its 1979 Yearbook to this important area of applications in school mathematics, which incidentally also had chapters discussing the issue of modelling. When the *Curriculum and Evaluation Standards for School Mathematics* was published by the NCTM in 1989, it highlighted one of the important goals for teaching mathematics as: "Mathematically literate workers" (p. 3). The document cited Henry Polak's comments about the mathematical expectations of new employees in industry:

- The ability to set up problems with the appropriate operations.
- Knowledge of a variety of techniques to approach and work on problems.
- Understanding of the underlying mathematical features of a problem.
- The ability to work with others on problems.
- The ability to see the applicability of mathematical ideas to common and complex problems.
- Preparation for open problem situations, since most real problems are not well formulated.
- Belief in the utility and value of mathematics. (p. 4)

Although, each of the seven points mentioned above has something to do with applications and modelling, the 1989 *Standards* document did not explicitly address the issue of applications and modelling nor did any of the subsequent *Standards* published by NCTM. As a result, there has not been the same focus on applications and modelling in school mathematics in curricula based on the ideas of the NTCM Standards. However, some international studies have been quite influential. The Programme for International Student Assessment (PISA) which is organised by the Organisation for Economic Co-operation and Development (OECD) countries also assesses an aspect of mathematical literacy not exactly similar but not entirely different as well to the one described above.

> Mathematical literacy is an individual's capacity to identify and understand the role that mathematics plays in the world, to make well-founded judgements and to use and engage with mathematics in ways that meet the needs of that individual's life as a constructive, concerned and reflective citizen. (See OECD, 2003, p. 24.)

The PISA Framework highlights mathematising as having the following five steps:

1. Starting with a problem situated in reality.
2. Organising it according to mathematical concepts.
3. Gradually trimming away the reality through processes such as making assumptions about which features of the problem are important, generalising and formalising.
4. Solving the mathematical problem.
5. Making sense of the mathematical solution in terms of the real situation. (p. 27)

The above steps connect nicely with approaches to applications and modelling. Furthermore, it should also be noted that the International Conferences on the Teaching of Mathematical Modelling and Applications (ICTMA) have also been very active with biennial conferences since 1983. The growing interest of educators in the area prompted the 14th International Commission on Mathematical Instruction (ICMI) study to be on the theme *Modelling and Applications in Mathematics Education*. As a whole, there is a growing number of resources for the school teacher to tap on, including resources from ICTMA.

Singapore has lived through two decades of a problem solving mathematics curriculum. The novelty of problem solving has waned off and its power to challenge and excite is not quite the same. Applications and modelling, offer new and exciting areas for explorations in the curriculum. Also, the long term benefits are quite good for students involved in the learning of these topics. The topics are fairly new to the Singapore teachers and students alike. The available textbooks at primary and secondary levels do look into aspects of applications however; they do not specifically address modelling. Furthermore, assessments and examinations in the local context also do not address these issues and as such the topics are marginalised. There needs to be a stronger emphasis by the authorities and more intensive workshops to get across the message. The mere inclusion of the topics in the framework is not enough. The topics should be somehow translated into teachable units for the teachers. The topics have to be incorporated in the regular curriculum. Textbook writers should be directed to include these topics in the textbooks that they propose for schools. Other resources such as calculators cannot be neglected as well. The Graphing Calculator (GC) has been introduced for many years now in other educational systems but it is only used at the Junior College (JC) (grades 11 and 12) level in Singapore since 2006. Although, there is a fair use of technology in the teaching of mathematics in schools, there is not much discussion about the potential of technology for modelling and applications.

Curriculum time is another issue that needs to be addressed. In an exam-oriented system like Singapore, the use of curriculum time for teaching about topics that may not have a direct impact on students' performance is difficult for schools and teachers to justify. There is no doubt that applications and modelling require more than the usual amount of curriculum time when compared with traditional topics in the curriculum. The benefits of applications and modelling may not be immediate and hence is a major handicap in their implementation when compared to some other topics. Like many mathematical topics at higher levels, applications and modelling can also trickle down to the lower levels. On the one hand applications and in particular modelling may require very high level of mathematics but on the other hand these topics can be honestly discussed at the lower levels as the work of some researchers has demonstrated (see English, 2007). Presently, benchmarking of school curricula against best practices elsewhere seems to be the norm. Although, we can speak of moderate successes in some cases, there is little to justify the place of applications and modelling in the school curriculum other than the lopsided views of mathematicians and mathematics educators who believe about their benefits in the long run for students.

5 The Final word

A discussion about the future of applications and modelling in mathematics requires some incursion into the history of mathematics. Bishop (1988) had claimed that in all cultures there was a need for counting, measuring, locating, designing, playing, and explaining which all required some knowledge of mathematics. Even before mathematics was organised into a coherent deductive system by the Greeks mathematics was deemed to be useful although it was empirical and passed on from one generation to the other. When the Greeks organised the knowledge about mathematics, more applications became apparent. Henceforth, mathematics was part of the intellectual and philosophical debate. Archimedes used mathematical ideas in his struggles with science. Later, Galileo was instrumental in placing mathematics at the pinnacle in the study of science. His famous experiments from the leaning tower of Pisa used mathematical calculations. His belief about the heliocentric system relied on his mathematics. Mathematics was used to predict the existence of stellar objects, such as planet Neptune, in the solar system. Einstein used mathematics to develop his theory of relativity and quantum mechanics. During the war mathematics was used to encode and decode messages. And the list goes on and on. However, different people looked at applications differently. There were those who did not believe or saw any future for the applications of mathematics. Hardy (1877-1947), the famous Cambridge mathematician, was one for example. He is supposed to have said in one of his toasts: "To pure mathematics and may it be of no use to any one." However, we know that we cannot go to such extremes as even the most unthinkable ideas that became part of the mathematical folklore have found applications.

Where do we see applications and modelling in the near future in school mathematics? There are three possible scenarios: (1) applications and modelling will die a slow death and will be ignored by most if not all curricula over the world; (2) applications and modelling be just another topic deemed worthwhile to consider but which otherwise may not be considered important; and (3) applications and modelling become the main principles around which the whole of the school mathematics curriculum would be structured. It is difficult to imagine the first scenario as mathematics is becoming increasingly important in the highly complex global world. The third scenario would be the best we can imagine, however, common sense, if nothing else tells us that will probably be not the case. The most probable is the second or somewhere between the second or third scenario. There is now a growing interest in applications and modelling in Singapore. The Association of Mathematics Educators rightly proposed Applications and Modelling as the major theme of the 2009 Mathematics Teachers Conference (MTC) which went a long way towards creating a greater awareness about the topics in the local context.

References

Bishop, A. J. (1988). *Mathematical enculturation: A cultural perspective on mathematical education*. Dordrecht, The Netherlands: Kluwer Academic Publishers.

Blum, W., Galbraith, P. L., Henn, H, & Niss, M. (Eds.). (2007). *Modelling and applications in mathematics education: 14th ICMI Study*. New York: Springer.

Dindyal, J. (2009). *Applications and modelling for the primary mathematics classroom*. Singapore: Pearson.

English, L. D. (2007). Complex systems in the elementary and middle school mathematics curriculum: A focus on modeling. *The Montana Mathematics Enthusiast, Monograph 3*, 139-156.

Ministry of Education. (2006a). *Mathematics Syllabus: Primary*. Singapore: Author.

Ministry of Education. (2006b). *Mathematics Syllabus: Secondary*. Singapore: Author.

National Council of Teachers of Mathematics. (1989). *Curriculum evaluation and standards for school mathematics*. Reston, VA: Author.

Niss, M., Blum, W., & Galbraith, P. L. (2007). Introduction. In W. Blum, P. L. Galbraith, H. Henn, & M. Niss (Eds.), *Modelling and applications in mathematics education: The 14th ICMI study* (pp. 1-32). New York: Springer.

Organisation for Economic Co-operation and Development. (2003). *The PISA assessment framework - mathematics, reading, science and problem solving knowledge and skills*. OECD.

Sharron, S., & Reys, R. E. (Eds.). (1979). *Applications in school mathematics: 1979 Yearbook*. Reston, VA: National Council of Teachers of Mathematics.

Contributing Authors

Judy ANDERSON is currently an Associate Professor of Mathematics Education in the Faculty of Education and Social Work at the University of Sydney. As the current President of the Australian Association of Mathematics Teachers (AAMT), she advocates for teachers of mathematics at all level of schooling and promotes the need for a quality mathematics curriculum. During her 33 years as a secondary school teacher and university lecturer, Judy has actively contributed to the professional development of primary and secondary school teachers. She coordinated the writing team that developed the mathematics syllabuses for Kindergarten to Year 10 in NSW in 2002. She conducts research into teachers' use of problem-solving approaches to learning and into students' engagement in the middle years of schooling. Judy is a member of the Editorial Board for the Australian Senior Mathematics Journal (ASMJ) and publishes in both research and teacher education journals.

Glenda ANTHONY is Professor of Mathematics Education at Massey University, New Zealand. Her primary research interests include mathematical inquiry and effective teaching practices and teacher education. She is the co-author of the New Zealand Iterative Best Evidence Synthesis (BES) for effective mathematics teaching and the *Effective Pedagogy in Mathematics* Educational Practice Series produced by the International Academy of Education. In addition, she has also published a number of papers in international journals such as Review of Educational Research, Contemporary Issues in Early Childhood, Asia-Pacific Journal of Teacher Education, and Journal of Mathematics Education.

Gayatri BALAKRISHNAN is a senior curriculum specialist in the Curriculum Planning and Development Division (CPDD), Ministry of Education, Singapore. Her work entails designing and developing the secondary mathematics syllabuses and supporting materials. She leads in the design and implementation of the additional mathematics syllabuses at the secondary level. Prior to her posting to CPDD, she was teaching in a secondary school for ten years. She holds a Masters degree in Mathematics.

CHAN Chun Ming Eric is a lecturer at the Mathematics and Mathematics Education Academic Group in National Institute of Education, Nanyang Technological University. He lectures on primary mathematics education in the pre-service and in-service programmes. His interests include children's mathematical problem solving, mathematical modelling, problem-based learning and learning difficulties.

Jaguthsing DINDYAL holds a PhD in mathematics education and is currently an Assistant Professor at the National Institute of Education in Singapore. He teaches mathematics education courses at both the primary and secondary levels to pre-service and in-service school teachers. His interests include geometry and proofs, algebraic thinking, international studies and the mathematics curriculum

Frank M. FÖRSTER is a Senior Lecturer at the Institute for Didactics of Mathematics and Elementary Mathematics of the Technical University of Braunschweig, Faculty of Humanities and Educational Sciences. He has a master's degree for mathematics and a master degree in teaching for lower and upper secondary level in physics, mathematics and pedagogy. After traineeship and work as secondary school teacher he returned to university as teacher educator. He has worked as a guest professor at several German universities (e.g. the University of the Saarland). His areas of research include teachers' cognition and beliefs, teaching mathematical modelling and applications in school and with gifted children.

Esther GOH Lung Eng is a Curriculum Planning Officer in the Curriculum Planning and Development Division (CPDD), Ministry of Education, Singapore. Her work involves designing and developing the secondary additional mathematics syllabuses and supporting materials. Prior to her posting to CPDD, she was teaching in a secondary school for eight years before moving on to do a Masters Programme in Curriculum and Teaching.

Roberta HUNTER is currently a Senior Lecturer and Co-ordinator the Graduate Teaching Pre-service Programme at Massey University. Her primary interest in research encompasses diverse students' participation and discourse in communities of mathematical inquiry. She is the author of several papers on the development of mathematical inquiry and argumentation with elementary aged students. She has served as a reviewer in various international journals including the International Journal of Science and Mathematics Education.

Gabriele KAISER holds a master's degree as a teacher for mathematics and humanities, which she completed at the University of Kassel in 1978. She has taught in school from 1979-2000. She completed her doctorate in mathematics education in 1986 on applications and modelling and her post-doctoral study (so-called 'Habilitation') in pedagogy on international comparative studies in 1997, both at the University of Kassel. Her post-doctoral study was supported by a grant of the German Research Society (DFG). From 1996 to1998 Gabriele Kaiser worked as a guest professor at the University of Potsdam. Since 1998, she is professor for mathematics education at the Faculty of Education of the University of Hamburg. Her areas of research include modelling and applications in school, international comparative studies, gender and cultural aspects in mathematics education and empirical research on teacher education. She has received grants from the German Research Society (DFG) in order to support this research. At present she is Editor-in-Chief of the journal "ZDM – The International Journal on Mathematics Education (formerly Zentralblatt fuer Didaktik der Mathematik), published by Springer Publishing House and Editor of a monograph series "Advances in Mathematics Education" published by

Springer Publishing House too. Since July 2007 she serves as president of the International Study Group for Mathematical Modelling and Applications (ICTMA), an ICMI affiliated Study Group.

Berinderjeet KAUR is an Associate Professor of Mathematics Education at the National Institute of Education in Singapore. She has a PhD in Mathematics Education from Monash University in Australia, a Master of Education from the University of Nottingham in UK and a Bachelor of Science from the University of Singapore. She began her career as a secondary school mathematics teacher. She taught in secondary schools for 8 years before joining the National Institute of Education in 1988. Since then, she has been actively involved in the education of mathematics teachers, and heads of mathematics departments. Her primary research interests are in the area of classroom pedagogy of mathematics teachers and comparative studies in mathematics education. She has been involved in numerous international studies of Mathematics Education. As the President of the Association of Mathematics Educators from 2004 - 2010, she has also been actively involved in the Professional Development of Mathematics Teachers in Singapore and is the founding chairperson of the Mathematics Teachers Conferences that started in 2005. On Singapore's 41st National Day in 2006 she was awarded the Public Administration Medal by the President of Singapore.

Marian KEMP is a Senior Lecturer at Murdoch University where she is the Head of the Student Learning Centre. She is a mathematics educator and her work involves supporting undergraduates in mathematics and statistics as well as developing programs for numeracy across the curriculum. Her research has been involved with the development of student strategies for interpreting graphs and tables. In 2007 she was awarded a Carrick Institute Award for University Teaching for outstanding contributions to student learning in the development of critical numeracy in tertiary curricula She has also worked with primary and secondary teachers throughout Australia and internationally in the context of developing such strategies for school pupils through the use of a *Five Step Framework*.

Barry KISSANE is a Senior Lecturer in Mathematics Education at Murdoch University, in Perth, Western Australia, having recently completed a term as Dean of the School. He has extensive background in mathematics curriculum and teaching, with principal interests in secondary school teaching. He is especially interested in the use of electronic technology for learning and teaching mathematics, having worked with various technologies, such as computer software and the Internet, but is most interested in hand-held technologies, such as graphics calculators. He has worked in both Australia and the United States in curriculum development and has presented many conference workshops and papers at professional conferences in Australia, many South-East Asian countries, and elsewhere. He has written some school textbooks, some graphics calculator books and many papers concerned with aspects of mathematics teaching. He has served as Executive Editor of the *Australian Mathematics Teacher* and as President of the Australian Association of Mathematics Teachers, among various professional roles.

Christoph LEDERICH is studying at the University of Hamburg for a master's degree. He is a prospective teacher for lower and upper secondary level mathematics and religious education. He will complete his study in 2011. Christoph Lederich is working in various projects on the teaching of mathematical modelling and on the professional knowledge of mathematics teachers.

NG Kit Ee Dawn is a lecturer with the Mathematics and Mathematics Education Academic Group at the National Institute of Education, Singapore. Her research interests include teaching and learning through contextualised tasks, interdisciplinary project work, applications and mathematical modelling at middle school levels. She is also currently preparing pre-service primary and secondary mathematics teachers for their teaching careers.

Verena RAU is studying at the University of Hamburg for a master's degree. She is a prospective teacher for lower and upper secondary level mathematics and physics. She will complete her study in 2011.

Verena Rau is working in various projects on the teaching of mathematical modelling.

Gloria STILLMAN is currently an Associate Professor of Mathematics Education at the Aquinas Campus (Ballarat) of the Australian Catholic University (ACU), Victoria, Australia. Her primary research interests include the teaching, learning and assessing of mathematical modelling and real world applications in secondary schools and in pre-service teacher education. She is the author of many papers and book chapters on modelling and applications which have appeared in scholarly and teacher journals. She is secretary and an elected member of the international executive of ICTMA (the International Community of Teachers of Mathematical modelling and Applications) as well as the Vice president (Research) of MERGA (The Mathematics Education Research Group of Australasia). She is a member of the editorial board *of ZDM—the International Journal on Mathematics Education* and a co-editor of *Australian Senior Mathematics Journal*.

TOH Tin Lam is an Assistant Professor with the Mathematics and Mathematics Education Academic Group, National Institute of Education, Nanyang Technological University, Singapore. He obtained his PhD in Mathematics (Henstock-stochastic integral) from the National University of Singapore. Dr Toh continues to do research in mathematics as well as in mathematics education. He has papers published in international scientific journals in both areas. Dr Toh has taught in junior college in Singapore and was head of the mathematics department at the junior college level before he joined the National Institute of Education.

YEN Yeen Peng is a senior curriculum specialist in the Curriculum Planning and Development Division (CPDD), Ministry of Education, Singapore. Her work entails designing and developing the secondary mathematics syllabuses and supporting materials. She leads in the implementation of the secondary mathematics curriculum to ensure that the content, pedagogy and assessment are aligned with curriculum policies and are well implemented in the schools. Prior to her posting to

CPDD, she was teaching in a junior college for ten years. She holds a Master degree in Mathematics Education.

Joseph Boon Wooi YEO (M. Ed.) is a lecturer with the Mathematics and Mathematics Education Academic Group, National Institute of Education, Nanyang Technological University, Singapore. He has a First Class Honours in Mathematics and a Distinction for his Postgraduate Diploma in Education. He has taught students from both government and independent schools for nine years and is currently teaching pre-service and in-service teachers. His interests include mathematical investigation, solving non-routine mathematical problems and puzzles, playing mathematical and logical games, alternative assessment, and the use of interesting stories, songs, video clips, comics, real-life examples and applications, and interactive computer software to engage students.

YEO Kai Kow Joseph is an Assistant Professor in the Mathematics and Mathematics Education Academic Group at the National Institute of Education, Nanyang Technological University. Presently, he is involved in training pre-service and in-service mathematics teachers at primary and secondary levels and has also conducted numerous professional development courses for teachers in Singapore. Before joining the National Institute of Education, he held the post of Vice Principal and Head of Mathematics Department in secondary schools. He has given numerous presentations at conferences held in the region as well as in various parts of the world. His publications appear in regional and international journals. He was part of the team at the Research and Evaluation Branch in the Singapore's Ministry of Education between 1998 and 2000. His research interests include mathematical problem solving in the primary and secondary levels, mathematics pedagogical content knowledge of teachers, mathematics teaching in primary schools and mathematics anxiety.